Sisters of the Wind

Sisters of the Wind

Voices of Early Women Aviators

Elizabeth S. Bell

Trilogy Books
Pasadena, California

Publisher's Cataloging in Publication

Bell, Elizabeth S.
Sisters of the wind: voices of early women aviators / Elizabeth S. Bell.
p. c.m.
Includes bibliographical references and index.
ISBN 0-9623879-4-0

1. Women air pilots--Biography. 2. Women air pilots--Biography --History and criticism. 3. Women in aeronautics. I. Title.

TL521.B45 1994 629.13'0922
93-061763

The author is grateful for permission to quote from previously published sources:

Excerpt from *Dancing at the Edge of the World* by Ursula LeGuin, 1989. Used with permission of Grove/Atlantic Monthly Press.

Excerpts from *Bring Me a Unicorn: Diaries and Letters of Anne Morrow Lindbergh, 1922-1928*, copyright 1972 by Anne Morrow Lindbergh, reprinted by permission of Harcourt Brace and Co.

Excerpt from *War Within and Without: Diaries and Letters, 1939-1944*, copyright 1980 by Anne Morrow Lindbergh, reprinted by permission of Harcourt Brace and Co.

Excerpts from *North to the Orient*, copyright 1935 and renewed 1962 by Anne Morrow Lindbergh, reprinted by permission of Harcourt Brace and Co.

Excerpts from *Listen! The Wind*, copyright 1938 and renewed 1965 by Anne Morrow Lindbergh, reprinted by permission of Harcourt Brace and Co.

Excerpts from *My Life*, reissued as *Alone in the Sky*, 1979, by Jean Batten, reprinted by permission of Airlife Publishing, Ltd.

This book is dedicated to women aviators,
past and present,
whose voices join the ongoing conversation.

Acknowledgements

Although writing is usually a solitary occupation, no one writes a book alone. Surrounding any writer are a multitude of people providing information, encouragement, support, and interest. We write as members of an extended community.

A great many people provided that community for me, and I take this opportunity to thank them. Tom Hobbs, Maria Sajwan, Bridget Smith, and Ann Wilkie of the Gregg-Graniteville Library at the University of South Carolina at Aiken worked miracles in locating and securing the sometimes obscure and out-of-print materials I needed for this project. Their expertise and help over the past two years have been invaluable. My university, also, provided help by granting me a one-semester sabbatical which allowed me to devote extended periods of time to research on these women aviators and to the fun of discovery that inevitably accompanies a project of this kind..

Also providing information with overwhelming graciousness were Joan Hrubec of the International women's Air and Space Museum in Centerville, Ohio, Dan Hagedoern and Larry Wilson of the National Air and Space Museum in Washington, DC, and Gloria Pitman of the University of California at San Jose. Brian Riddle of the Royal Aeronautical Society in London, Kathy Farrett of Quadrant Picture Library in Surrey, and Joe Canale of Associated Press/Worldwide Pictures kindly allowed me to select from their archives the photographs that accompany this volume.

Other people offered valuable intangible support. Phebe Davidson, Sandy Hochel, Sue Lorch, Linda Owens-Whitlaw, and Charmaine Wilson listened when I needed them to and provided encouragement at every step of the way. Lydia Bell and Jason Bell were always willing to give advice, and they balanced that advice with an unswerving faith in the project. Finally, Ron Bell, my research partner, freely gave his time effort, research expertise, friendship and comfort, even as he maintained a healthy sense of humor about the entire project.

To say thank you to my community of support seems pale beside the gifts they have given me. Nevertheless, I extend my heartfelt thanks to them all.

Contents

Introduction

Carolyn Heilbrun in *Writing A Woman's Life* (1988) charges that because of societal pressures women historically have been unable to tell the truth of their life stories, having instead to either romanticize or submerge their perspectives in order to be accepted as real voices. In fact, she says, no true account of their lives *could* have been written by women before our own last half of the twentieth century; during our time women have seized the strength to express anger and frustration, to demand recognition for the full scope of our humanity. Heilbrun's words ring true, to be sure, but they do not fulfill. They cannot empower because they focus on the void, on that which could not be told. While women's life stories surely have grown in inhospitable ground, they have been hardy. They contain, as much in their spaces as in their statements, truths about their authors' worlds. They have preserved the nuances—the positive ones as well as the negatives—of women's lives. To ignore that is to violate even further the body of women's experience. But, even accepting that our own perspective, which has been forged in the vocal and activist latter half of the twentieth century, cannot fully comprehend the subtlety of earlier women's autobiographies, we must still ask, What do those women's autobiographies record? What do they tell us about women in general or about that one particular woman? In short, what do they tell us that we need to know?

One of the problems involved in exploring the range of what women found possible to say about themselves has been that their lives have taken place in different contexts. Historically only a few women chose to record their experiences. The life of a Margery Kempe, the 15th century mystic, differs in crucial ways from that of a Mary Rowlandson, the 17th century colonial who wrote of her captivity at the hands of Indians. And they both differ radically from the life of a contemporary Annie Dillard or Anne Sexton. Therefore, they write of different subjects and create variations in their narratives which one can attribute to a multitude of reasons. While this certainly true and valuable patchwork convinces us of the scope of women's lives, it leaves unanswered a more difficult question: What is the range of women's reactions to similar experiences? Would we, for example, find any similarity in Kempe's and Dillard's perceptions of women's lives if we could eliminate the centuries and provide them with similar circumstances? Or to put the question in Heilbrun's frame: Do women share a core of experiences that defy time and place to belong inexorably to women as a group? That latter question brought me to the writers included in this volume.

Here in the opening decades of the twentieth century, in an island of time surrounded by two world wars, we have the writings of a group of women engaged in one of the most significant enterprises of our age. As aviators during the infancy of human flight, they explored, set records, expanded human potential, and risked their very lives for a dream they wanted to make a concrete practical reality. More to the point, they wrote about it, recording in their own words what they found worth telling about. This handful of texts represents a goldmine of information about the range of women's reactions to similar experiences. It provides vital pages of our national and cultural heritage, not just women's heritage—although it does give us back part of our knowledge of that—but of our heritage as a people, a culture of human beings.

While beginning preliminary research several summers ago for my original project on early aviators as a group, I discovered that the majority of texts written by male aviators of the relevant time period (1920-1940) were readily available, carefully archived; many of those written by women aviators were not. In fact, material even listing the writings of women aviators proved to be virtually inaccessible. To further complicate the matter, I discovered that many archives, also separate their holdings on aviation into two categories: "Aviators," which holds the material on male aviators, and "Women Aviators," which holds the material on female aviators. By definition, then, the work of women is often treated as a subcategory of the whole or, perhaps more damaging, a category not belonging to the "real" or whole subject. In subsequent dealings with rare and out-of-print book dealers, I discovered that while I could find multiple copies of the male aviators' autobiographies, I found virtually none of those by the women. Resorting to special collections in various libraries across the nation produced mixed results: Copies of some of the autobiographical writings of British and American women aviators are nonexistent in this country; they just simply disappeared from any kind of public forum. Two or three copies of other women aviator's autobiographies exist in closed collections; access to them is extremely limited. Others are more readily available for scholars and interested readers, and some are even being recognized anew as valuable pieces of writing, as the recent reissue of Amelia Earhart's books indicates. But certainly, the reader who wants to read all or most of these autobiographical texts will have great difficulty in even locating them, much less obtaining them.

Furthermore, I found that many people no longer remember the names of women flyers from five or six decades ago, not even those who were in the center of media attention. Some of those people may have heard of the aviation exploits of Charles Lindbergh, Richard Byrd, and Antoine de St. Exupéry, but many of them do not know that author Anne Lindbergh was also a licensed pilot and radio-operator/navigator for all the commercial air-route-mapping flights she and Charles made in the 1930s. Or that aviator Amelia Earhart wrote three nonfiction books and a regular aviation column for *Cosmopolitan*, as well as other articles for a variety of magazines. Many of the other women who set records and worked extensively in establishing what we now recognize as the commercial aviation business have been dropped from the public consciousness and their written accounts—those which survive—have been relegated to forgotten shelves. The Louise Thadens, Mary Bruces, Amy Johnsons, Jean Battens, and the others have much to tell us.

Their accounts, whether official autobiographies or, in Anne Lindbergh's case, also published diaries, record the experiences of a remarkable group of women who entered a field not only unconventional for women but also in some places actively hostile. These women faced daily the tremors of a society unequipped to see them in their roles as pilots, businesswomen, navigators, daredevils, or record holders. The most articulate of these women found at the height of their fame immediate markets for autobiographical writings about their experiences, for the public still saw women's accomplishments of any kind as evidence of a superior woman and, therefore, somewhat of a spectacle. Yet, the women aviators wrote for different purposes than the popular media coverage of them: They saw themselves as pioneers, and through their writing, these women could voice their personal dreams, their concerns about societal expectations, and their views on the future of aviation, as well as recount their own endeavors with a degree of insight and humor that makes fascinating and remarkably timely reading. Although their writings bear the strong and unmistakable imprint of their individual personalities and circumstances, they of necessity address many similar issues.

In their written accounts, virtually all these women aviators responded to two levels of myth society wove around them. First, these women were very aware that they were participating in humanity's grand adventure into the skies, breaking gravity's "absolute" grip on our dreams. Hailed as the promise of a new era in human achievement, the advent of aviation led the popular press to launch a flamboyant campaign comparing flyers to legendary heroes such as Daedalus, Icarus, Prometheus, and other gods of classical mythology. The media focused on, of course, the men who flew, but even as the women began to make their marks, the press continued to use the same allusions, often granting to the women the role of "lady god," sometimes by an even more submerged turn of phrase. Amelia Earhart, for example, found herself labeled "Lady Lindy" after Charles Lindbergh, and while it conferred on her many of the same glories it had on him, she still chafed at the secondhand nature of it, as did Amy Johnson, known as "England's Lindbergh."

The autobiographical writings in this study show an industry growing to age in a new world, but more importantly they reveal women of courage and spirit defying the expectations of a reluctant society in order to find and communicate a growing sense of self value. Each woman selected for her readers the ideas she felt most accurately conveyed a sense of "what it was like" to have been a principle participant in the drama of the age. These women tell us much we need to remember, but we need also to recognize the parameters of what they tell us. Thus, some will say of this volume, "But it doesn't present the real Thaden, or Bruce, or Earhart, or Lindbergh. . ." as if such a person *could* be portrayed in the medium of black and white print. Indeed, two of the most recent and thoroughly researched biographies of Amelia Earhart (Mary S. Lovell's *The Sound of Wings* and Doris L. Rich's *Amelia Earhart: A Biography*) contradict each other in a multitude of minor and major conclusions about the "real" Earhart. And how many versions do we have of Anne Morrow Lindbergh, each produced by careful and impassioned scholars. No, I don't propose in this volume to present the "real" anyone, for I believe the biographer's art to be too inexact to promise that. Rather, I propose something vastly different: this volume presents its subjects in the guise they chose themselves, at a particular juncture of time and place, as the way they wanted their contemporaries to see them. There are several qualifiers to this proposal, but in their parameters, we must find a particular kind of reality, a reality born of the author's perception of her times and of her place within them. Within their narratives we find embedded the shape and form of the society which produced these women; within the compliance or rebellion they displayed for these shapes and forms, we find their personalities.

By virtue of living within a society, we absorb a knowledge of its guidelines and its etiquettes. Such knowledge comes to us unbidden, subconsciously through the behaviors we see modeled around us and the reactions these behaviors elicit. We cannot escape such knowledge, for it contributes to the daily commerce of our age. So, too, did our authors see their world. Thus, they "knew" with that subliminal, visceral certainty, the nature of the society they inhabited. Just as surely, they knew as well how comfortable—or not—they were within its boundaries. Each of our authors stepped beyond the limits she knew her society set for women, but also in so doing, each decided, consciously or not, how deliberately she wanted to challenge these boundaries. The result is that each author spoke to her contemporaries in the voice and manner she felt best communicated her story, and, consequently, best described her relation to her contemporary world. She tells us a great deal about what she valued, for her values permeate the choices of subject, voice, narrative approach, and resolution she adopted. Her audience, those real, if anonymous, faces she saw on the street and over the banquet table, those people who shared her time and place, served to guide her in these decisions.

And here we find an interesting juxtaposition. The world for which our authors wrote was a far different place than the world from which we read them. In a very literal sense, they did not write *to* us, although in truth they may also have written *for* us, to record for future generations their passage in the world. But for our authors those future generations—those generations that have become us—did not yet exist; they had no faces, no substance, no expectations, no etiquettes. In the absence of a concrete reality, those future generations could not contribute to the authors' understanding of what they wanted and needed to say. Inescapably, then, in our authors' narratives, we read a message intended for someone else; the meanings we derive tell us as much—if not more—about ourselves and our societal values than they do about our authors.

Curiously, however, the very nature of the distortion passing time has created in our understanding of these narratives ensures their authors a form of longevity, for in reading of their lives and their choices and their victories and their discoveries, we invite these women into our own world. In recognizing the extent of their achievement, we validate (yes, and celebrate) their lives anew. If the voices with which we hear them speak to us sound as if they issue from a victrola, they also sound courageous and full of the spirit of life which time can't erase. But even more importantly, each woman speaks in a voice particularly her own, full of the nuances she felt as an "uncommon" woman in her world. Each gives us a sense of her "self" as vivid and as substantial as if she stood before us. Each, too, picks up resonances from her sister aviators. Their voices, each distinct, blend to create a conversation about what those early years of aviation and those still-impressive adventures taught them, what they gained, and what they wanted to leave those of us living in the future world they helped to shape. Through their words we follow their progress, individually and collectively, and through their journeys we learn of society's as well.

The selection process for authors to include in this study forced some difficult decisions, however. My purpose has been to show women as they helped to shape the course of modern aviation in the age when it was still exploring its possibilities. Only two decades fit that description—the 1920s and the 1930s—when aviation was old enough to be seen as more than a toy, but young enough to need to prove its value in the real world. I wanted also to avoid the war years at both ends of the scale because military uses of aviation differ radically from everyday commercial ones. Those daily routine uses of flight to which women aviators made major contributions have become so commonplace that a brief reminder of the dangerous and frustrating process that led to them seems invaluable.

But my interest has always been divided between the aviation achievements *per se* of these women aviators and their written accounts of them. I decided to examine them both, but only in terms set out by the women themselves. I decided to include, then, only women aviators who wrote autobiographical accounts of the flying they did in the 1920s and 1930s, knowing full well that I was thereby ignoring the accomplishments of many skilled aviators who did not write about themselves—women such as Bessie Coleman, the first African-American woman to gain a pilot's license, or Phoebe Omlie, who earned numerous records, or Ruth Elder, who became a media sensation and aviation's "bad girl." Furthermore, my decision to limit my selection to those women who actually published their autobiographies in the 1920s or 1930s eliminated other authors, such as Ruth Nichols who worked for the access of women to training and employment in aviation but who published *Wings for Life* in 1957, or even Beryl Markham who published *West with the Night* in 1942.

While this decision may seem arbitrary, it rests on a philosophical foundation. I wanted to read accounts written when the experience was still fresh. Perspectives change over the years, especially over years as dramatic as the 1940s proved to be, and while experience recounted in retrospect provides valuable insight, it inevitably is different insight than one gains from more immediate accounts. I wanted to hear the voice of the woman who flew those flights, not the voice of the woman she became, especially after the Second World War irrevocably changed the world. And so, I left out many women who also accomplished wonders and who also need to be heard, but in a context more fitting to their circumstance.

Strangely, the most difficult decision I made about the selection of authors to include involved the passenger accounts with which I begin this collection. They seemed a different order of account than the pilots could create. Ultimately, however, I decided that the narratives of women travelers who had not and never would pilot a plane offered a valuable reminder that not so long ago women were considered such limited creatures even their *traveling* in an airplane was newsworthy. They, the not-so-famous, also charted new roads for women, roads most of us now frequent with such nonchalance that we can't imagine the daring of those first women passengers. Their stories need to be heard again, for they tell us much about our gendered past, whether we are female or male. In looking at the absurd stereotypes these women faced, perhaps we may learn to view our own in a more knowledgeable light.

Finally, I decided to maintain the spellings of place names and the national boundaries as they were recorded in the writings of our authors.

Thus, the world I discuss is a world without Pakistan or Bangladesh with Karachi still in India, a world in which Vietnam is French Indochina and Persia governs large parts of the Near East. Hawaii and Alaska are still territories, and distances and speeds have altogether different connotations than they do today. It is the world our women aviators knew well, for they had visited its corners and covered its expanses.

That brings us back to the women, which is fitting, for the truth beneath this project, the one that serves as its guiding principle, resides with the women themselves. To a person, they were extraordinary women in extraordinary times doing extraordinary things. They deserve to be remembered.

1

The Passengers

While it may seem odd to begin a discussion of women who flew airplanes with a chapter on women whose claim to fame is that they flew *in* airplanes, such a beginning brings a needed balance to the picture. None of these women saw aviation as centerpieces of their lives. They, in fact, encountered airplanes only rarely and in the midst of other more pressing activities. Thus, they bring an objectivity and distance to the topic our aviator/authors cannot. They complete the picture of what air travel meant to women in the early decades of the twentieth century because they show the more public face of aviation enterprise. They show us what aviation was like for the uninitiated woman brushing up against the predominantly male world of the hangar and the airplane. After all, aviation had not been designed with women in mind. In fact, aviation had not been designed with passengers in mind; the concept of *women* air passengers boggled the mind. These passenger narratives tell us much about the accommodations the infant commercial industry made as it explored the feasibility of carrying people—even women people—from place to place in relative safety, comfort, and convenience.

In the 1920s, aviation was, after all, a new enterprise and one that had demonstrated itself to be dangerous in some very spectacular ways. The very activity of traveling in an airplane, especially in a world not yet peppered with airfields or travel schedules, proved adventure enough for most people, male as well as female. Early passengers rather inevitably found themselves with stories to tell, and their passenger narratives capture a more naive, impetuous phase of aviation history, even as they provide exciting reading. In fact, they fit very well a public reading taste of the time. The attraction of travel writing which had reached fevered pitch in the late 1800s had not yet abated; aviation passengers added a new element to the genre. They combined stories of exotic places and activities, the stuff of travel writing, with a sense of historical significance that went beyond the personal story of the "ordinary" travel writer. Aviation writers represented a new movement, a collective leap into a new era vastly different from anything previously experienced. As passengers, these writers presented the common person with the possibility of participation in that movement. Both the writers and their readers understood that they were sharing in the emerging myth surrounding aviation, that even though the passengers recorded more adventure than the average person could manage to accrue, the adventure of air travel might someday be achievable for ordinary people.

But equally as important, the advent of aviation represented but one more facet of a dramatically and rapidly changing world. Old patterns of thinking and acting no longer seemed appropriate, but frustratingly the new patterns were emerging only slowly and mainly in response to the tensions that necessarily arise when old paradigms break apart. Some of the strongest reverberations of this process, as might be expected, attended to gender roles. And that brings us to the crux of the matter and the partial explanation for a chapter on the memoirs of aviation passengers, virtually by definition a female genre.

By the 1920s, societal conventions proscribing appropriate behavior for women were being challenged on a number of fronts, from sexual behaviors to fashion, from legal definitions to careers. Old-fashioned propriety found itself confronted inevitably with new behaviors and new moral definitions. Even the emerging questions were new. Propriety, at least in its former guise, was falling by the wayside. A look at the history of aviation provides a capsule version of what was happening in society at large. Conceived of and recorded as virtually solely a male endeavor—after all, the major public contact with it so far had been with legends of the WWI flying ace who was now barnstorming and stunt flying to make a living—aviation appeared too demanding and too dangerous for women to participate in actively. Even so, in an effort to convince the public that air travel had practical applications, the aviation industry eagerly assigned women to the supporting role as their men considered careers as pilots, mechanics, airplane designers, and industry entrepreneurs.

Even many women writers, including those excited by aviation in the early 1920s, perceived women's role in aviation as predominantly a passive one. As an example, Daisy Elizabeth Ball, herself a member of the Aero Club and a journalist for *Aircraft*, writing for *Aeronautical Digest* in 1923 recognized that women such as Harriet Quimby actually became pilots, but she found the more valuable contribution women could make to aviation in Katherine Wright's devotion and support of her brothers, Orville and Wilbur. She perceived women as the mothers, wives and sisters of aviators, but not usually as aviators themselves. Likewise, in 1924 Margaret Davies wrote for *Aero Digest* that women must become a moral voice in aviation. Drawing on the image of women as nurturers and caretakers, Davies called for women to support national agendas for airpower as a preventative of war. Having seen first hand the horrors of air strikes on defenseless civilian women and children in war-torn London, she urged women to use their influence as mothers to protect future generations by making war inconceivable. Other voices joined theirs, emphasizing the traditional role of women behind the scenes or as spokespersons for the emerging aviation enterprise, but often they fell short of urging women to fly themselves.

Hence, women with a more than theoretical relationship to an airplane, those who actually set foot in a real airplane, found themselves, at best, curiosities in the aviation world, a world that neither wanted them nor welcomed them in any other capacity. To understand the complete narrative of aviation history we must examine the dimension their stories and their voices can add.

Gertrude Bacon, whose *Memories of Land and Sea* was published in 1928, presents the most troubling of passenger narratives, for she embodies strong contradictions between her actual activities and her description of them. In her 50s when her book was written, she belongs to an earlier generation than most of the women in this study. Her memoir takes place in retrospect, as its title implies, but in that retrospective Bacon captures the birth of modern aviation, a birth she witnessed at very close range. She creates a pastiche of memory, nevertheless wondering aloud if she has the right to do so, for she tells us two worries that have caused her to reconsider the wisdom of recording her story. First, she wonders if her age allows her enough distance from her readers to provide them with anything more than "common ground" (1) to read. In addition, she asks if her life could interest others enough to merit its preservation. We sense a ploy: Ultimately she, of course, chooses to write her book as a record of the many firsts in which she participated as history was being made, but as a reminder of her humility, she softens her role in her own life, she shifts the focus from her own activities to those of her companions, she steps back from the spotlight even as she subtly controls her narrative. Possibly as a result of her place in time, or perhaps of her particular family circumstance, Bacon blends her zeal for adventure with the remnants of a Victorian concept of propriety, thereby creating for today's reader a discomforting dissonance: For such an adventurous woman who engaged in so many unconventional activities for women, she portrays herself as curiously passive, as primarily the lucky observer to the achievements of a dazzling array of supermen, including her own father. Listen to her voice: "And again, though taking no personal share in the great deeds of the world, it has been my fortune to be not altogether remote from some of them" (1).

In fact, she addresses us through a complex dual persona, beginning the book with the submissive voice of a Victorian sage's dutiful daughter, ending it with the self-deprecatory tone of the older woman who would natter on too long if she's not careful. But she maintains throughout the book a confidence and self-assuredness that belies either guise. The complexity seems deliberate, the two dominant facets of her persona carefully chosen to embrace the sensibilities of her contemporary audience, for as she tells us in her memoirs, the connection she establishes with her audience determines the success or the failure of her work. Her overlay of confidence stems from her recognition that she knows, from experience, how to establish the bond she wants with listeners/readers, for she understands the nature of audiences.

Her discussion of audience comes in the latter part of her book, in a chapter dealing with her career as lecturer. Although the terms she uses in her analysis relate to speakers and audience, her insights translate easily into recognizable form for the writer as well. She labels the "subtle sympathy" between speaker and audience as one of the joys of lecturing. Its presence lifts the experience of lecturing to a level of personal fulfillment; its absence makes the task leaden. Given its relative importance, Bacon tells us, "At the beginning of a lecture the speaker has to establish this contact, and the ease with which he can do this varies largely according to the locality" (200). Within this framework, Bacon establishes two provisions of her theory on audience: One must make connection early with one's audience, and one must know the expectations and peculiarities of that collective audience. She describes in anecdotal form her experiences over the years with a wide variety of audiences, from those in mental hospitals and prisons to those in community organizations and celebrations.

Thus, I suggest, when writing the memoirs of her rather extraordinary life, Bacon established a persona and an approach with which she believed her contemporaries could empathize. Full of traditional values such as filial devotion and echoes of familiar female roles such as the nonthreatening older woman, Bacon's persona allowed her to describe for her audiences herself as a woman living an adventurous life without offending or alienating her readers. She selected her persona to fit what she believed her contemporaries would accept; she would, I believe, be among the first to admit that other audiences, other generations, might not find her choice so appealing.

Nevertheless, Bacon seems to legitimately accept some of the assumptions her society held about the nature of men's and women's appropriate roles. Thus, although she has a fondness for adventure, born perhaps in the reading she did as a child, she sees adventure as mainly a male prerogative. While she flies in balloons, descends into coal mines, inspects submarines, flies in aeroplanes, she does so only at a man's invitation, for these are elements of a masculine world. Over and over she uses such words as "I was chosen," "he gave me the chance," "I accompanied," "I helped." She does not initiate her adventures; she merely responds to the opportunity. But those opportunities arise with rewarding frequency.

Her childhood seems witness to a parade of famous old men, come to share their stories of adventure with her famous father. The stories always belonged to the men; infrequently women accompanied them, but usually only as wives. Bacon tells of one man, a famous architect, whose "charming invalid wife" lived a life of forced inactivity "compiling charming books of literary extracts" and who called herself "An old woman with a basket gathering treasures" (12). Bacon found the contrast too great to be ignored: In her childhood world, men's lives equaled adventure, women's were "invalid." One wonders if Bacon saw the pun contained in the word.

The first half of her book, in fact, consists of anecdotes culled from the stories those famous men told. Their names pepper her pages, along with the accolades she heaps on them, for they were revered by the world and not the least by Bacon herself. Only two or three times does she mention women of accomplishment, and then apparently grudgingly. She mentions as observers of a total eclipse she was photographing with her father "renowned astronomer and writer" Miss E. Brown and Mlle Klumpke of the Paris Observatory, but tells us nothing more of them. In another section of the book she mentions Mrs. Maurice Hewlett as a skilled aviator, but spends more time telling us Mrs. Hewlett had the pleasure of teaching her son to fly than in discussing her achievements. In fact, Bacon seems vaguely ill at ease with accomplished women: "Then there were rather terrifying lady astronomers from Vassar, the woman's university (why are clever women in general so much more alarming than clever men?)" (97). One possible answer to her question may lie in her relationship to her father.

By her account nothing short of a Victorian sage, Bacon's father held a central position in her life. Educating his children at home under his strict supervision, he controlled what they learned. Bacon even tells us that he selected her reading material, although she refers to it as "guiding her choice" of books (20). In addition, he channeled his growing children into his scientific projects, a practice that absorbed Bacon's adult life until her father's death in the early 1900s. Indeed, she was his assistant and even learned photography in order to be more useful to him. Her adventures with ballooning developed from her association with her father's attempts to use the new technology to expand his experiments. Bacon had little opportunity, if we take her account to be accurate, to see clever women at all. For her, this was literally a man's world. Ironically, of course, Bacon herself was a clever woman, a fact of which she never seems cognizant. In her book, she recognizes herself as female, she knows she is clever, but she does not define herself as a "clever woman." Instead, she identifies rather completely with her role as her father's daughter.

Bacon idealized, maybe even idolized, her father. She credits him with "setting the tide" of scientific uses of ballooning, and perhaps of aviation as an enterprise, in England. She assures us his name is well known, that he is well respected, and that he is "parent, teacher, comrade, guide, philosopher, and friend" (14) all in one. Until his death, her story sounds remarkably like a paean to a perfect father, and in her mention that after his death she wrote his biography, one hears the paean continue. Modern readers, however, look at her story in quite a different way. Again, we read with a dissonance her praise—adoration—of the father who "scolded, not petted, when we had colds or were otherwise ill" (13), who taught her that little girls should be seen and not heard, who said "Every boy should be taught two things" but "in his individual case `boy' meant `girl' also" (194), who at a gathering of astronomers viewing a total eclipse publicly admonished the adult Gertrude as "you silly child" (74), who commandeered her life as his due, even if it were with her own compliance. Modern readers look for what surely must be a touch of anger, irony, or sarcasm in her mentions of the above, but none appears. With apparent sincerity, Bacon tells of her father's skill and wisdom in educating her and his generosity in allowing her to work with him—her parent, teacher, comrade, guide, philosopher, and friend.

But she lived in a different age, an age rooted in the Victorian mindset. To a large extent, Bacon never shakes loose the enveloping folds of Victorian England. She accepts without question many of the prejudices and categories prevalent in Victorian England. These prejudices show most forcefully in Bacon's discussions of people, especially groups of people. For example, she looks with horror at the poverty and filth of colonial India, a "nightmare world" with "natives" of whom she expects nothing more. In this, she demonstrates the colonial attitude prevalent in England during what we now know were the waning days of the Empire (88). Of a visit to the United States, she credits her tolerance of African-Americans to her not having to live in proximity to them for long stretches of time. Otherwise, she is sure she would find them, just as the stereotype predicted, "slovenly" (96). Again looking at a group, Bacon frequently condemns the moral fiber, education, and decency of the younger generation. They just quite simply cannot measure up to the standards set by her own generation, especially as taught by her father. She recounts these beliefs with full confidence that her audience would understand and agree. Her prejudices, some virulent enough to be considered by today's audience as racism, existed toward groups of people, not people she saw as individuals. Nevertheless, they drew on negative stereotypes and resulted in Bacon's tacit dismissal of the value of the groups she so labeled. While as modern readers we deplore her views, we must also recognize that they were typical of her era. A question we must repeatedly ask ourselves is whether we can blame the writer for not stepping beyond the world view of her day.

Bacon complicates the matter for us, for on one level she understands the discrepancies in the values of different generations. Specifically, she understands that Victorian women related to their world in a much less open way than contemporary women, but finds in that difference a strength in the Victorian approach. "The Victorian ladies knew perfectly well—none better—what they were about. . .and they succeeded at least as well as their grand-daughters, with Eton crops, and no clothes and no manners worth mentioning; in fact, they would appear to have succeeded better. But underneath that studied pose which left preening grandfather feeling so much cleverer, wiser, stronger, and therefore susceptible, there lay a strength of mind and body and brain and spirit all the more potent because so rarely manifested" (206). Again, as with her discussion of her father, Bacon seems sincere in her assessment of the virtues of the sensibility of the past, and she seems totally convinced of its successful application in women's lives. But the modern reader catches her in a paradox, for Bacon also notes the limitations society placed on women's lives. In fact, these notices permeate her book.

She reminisces with a touch of humor at the absurd stigma she encountered when in 1894 she became one of the first women to ride a bicycle, in skirts five yards around at the hem, which was the acceptable attire. She laughs at having had to assure relatives and friends at the time that she was not the woman bicyclist the news media reported as having been forced to ride in the baggage car of a train because she wore trousers. She even gently ridicules the learned opinion of medical doctors at the turn of the century who warned of the dangers a bicycle seat posed for women (24-25). But while she sees the absurdity, she does not question it. While she laughs at those outmoded conventions that from her vantage point in 1928 seem so silly, she does not apply that insight to more current situations. Without comment concerning the inequity, she tells of watching women work at a coal mine, carrying huge pans of muddy sludge while under strict injunctions to keep their white aprons clean (122), as if clean aprons mattered in a coal mine or miners were required to keep themselves clean and neat. Sadly, the modern reader knows that even the presence of women workers at coal mines in the early 1900s, much less the unrealistic, absurd conditions under which they labored, virtually disappeared from the historical record.

In her own experience, Bacon finds herself discriminated against because she is a woman. She is refused passage on a publicity flight for reporters, all male, on the first commercial passenger flight across the English Channel. She also hears herself denied a listing with the London Lecture Agency because the agency's head believes all women lecturers are, by definition, "cranks" (195). Finally, she tells us, she is "allowed" to fulfill "most" of the lecture commitments left empty at her father's death (195). Ironically, however, she has also profited from this kind of prohibition, for her journalistic career was rooted in "first woman" accounts. Bacon never addresses the paradoxes inherent in her views; perhaps she never even realized they existed. Or perhaps she realized that she succeeded, in part, *because* of the prohibitions society placed on women in general.

Despite these inconsistencies in her vision and even despite the modern reader's discomfort with many of them, Bacon nevertheless writes with power and grace of the birth of the aviation age, and about that the reader in both Bacon's time and ours finds no hesitancy or doubt. She makes believers of us all. Bacon's experience in aviation encompasses balloons, zeppelins, hydro-planes, and various models of very early aircraft of all kinds. While she describes her connection to these vehicles passively, her knowledge of them commemorates a far more active encounter with them, derived from the journalistic work that took her into hangars as well as factories, submarines, coal mines, and other atypical places. Although she does not pilot them, she understands the mechanics and the aeronautical theory behind each mode of flight she encountered; she made it her business to learn the technicalities. Indeed, she discusses the relative advantages, or disadvantages, of various types of engines, explains how they work, gives the reasons for why engine placement or seating or fuselage shape differs from one model or design to another. She discusses speed (in her days a misnomer for few civilian planes in the 1910s and 1920s could surpass 100 mph) and altitude (at a time when scientists still discussed what might happen to a plane that traveled higher than 1000 feet), as well as the stresses flight places on pilots, even to knowing that certain kinds of engines spewed out a thin film of oil that could coat the pilot's eyes and distort his vision (171).

Assuredly an intelligent woman, Bacon also demonstrates a strong (dare I say, masculine) technical familiarity with aircraft of all kinds. She displays it matter-of-factly, doing nothing exhibitionary about calling the reader's attention to the information one would not, at the time, expect women to have.

Bacon is far from matter-of-fact, however, when she discusses for her reader what flying is like. Of her own reactions, Bacon's favorite descriptions involve some form of the word rapture. In passages almost poetic in their purity of language, she takes her reader into the air with her, pointing out the breath- taking beauty of the world above the clouds and the smoothness of the airplane's movement in the air. Flying remained a thrilling experience for her, in memory perhaps more strongly than in fact. She reminds us that early airplanes ignored the comfort, if not the actual presence, of the passenger. In several anecdotes she illustrates this with accounts of sitting next to the very hot engine, or sitting behind the engine that spews out fine mists of oil into her face, or having to have her weight checked to be sure she is not heavy enough to crash the machine, or scrambling into a plane with no steps for the passenger to mount. Clearly, Bacon confronts the world of aviation in its infancy. Despite the discomfort of the accommodations, however, she finds wonder and thrill in the adventure of flying, in "that glorious gliding sense that the sea-bird has known this million years, and which man so long and so vainly has envied" (162). She describes for her reader the sights one encounters above the clouds and the sensations of looping the loop, as well as that indescribable moment when the airplane lifts from the ground and begins to soar. Indeed, she serves as interpreter for her reader of experiences they have not yet encountered and perhaps never will.

Nevertheless, perhaps Bacon's most valuable contribution to aviation literature is her awareness of the historical place she held: "I have experienced something that can never be yours and can never be taken away from me—the rapture, the glory and the glamour of `the very beginning'" (162). She recounts the early nonchalance with which she heard family friend Octave Chanute, later to be called "the Father of Aviation," tell in 1903 of two American brothers and their seaside experiments with heavier-than-air flying machines. In fact, she tells us that only several years later did anyone really understand the importance of what the Wright Brothers had accomplished. Of her own first flight, she began to recognize its significance when, after it was over, people "congratulated me, begged my autograph, and toasted me in the wine of the Province; for they said I was the first Englishwoman to fly" (164).

And so began her solo career as premier writer and lecturer about aviation matters. Indeed, her life became a round of visits to "aerodromes," interviews with flyers, attendance at flying meetings, research from books, and research from personal experience which she recorded in books and articles for a public hungry to know more about this strange activity called flying: "and very happily I was borne along on the great wave of aerial enthusiasm and adventure that was then sweeping over the world" (165). Was her message, so lovingly and carefully recorded for her readers and for us, a valuable one? Do we find it as germane today as her readers did almost 70 years ago? The answer is yes, but for very different reasons than her readers would have recognized at the time.

In order to appreciate Bacon's contribution, we must also look carefully at what she does not tell us about aviation. She does not, for example, write of aviation safety. She does not try to convince her readers that flying provides a safe way of traveling. To try to do so would be perceived as patently foolish by her contemporary readers, for media records with which everyone was familiar easily disproved the very idea. Nor does Bacon write of long journeys, of flights leaving from one location and covering great distance to a destination for which one had a specific plan. Instead, she records flights for their own sakes, flights around the airfield or, in one case, around a lake. For most of Bacon's experience, especially the pre-war flights, aviation was still a toy; we were still exploring what we could expect from flight as a concept, for it was so minimally a reality. Early airplanes with which she was familiar lacked the horsepower to carry cargo; the open cockpits, minimal fuselages, and fabric-covered wings discouraged the uninitiated. Early aviators bartered with fate each time they flew. Bacon's omissions of what we now consider defining characteristics of aviation convey to the modern reader as much about that early world of aviation as her descriptions of it.

But Bacon's experience and her career forced her into a role for which we still find need. For Bacon's unique place in history cast her as an aviation philosopher as much as a practitioner. In exploring why aviation was not born until the twentieth century, she looks at human frailty; in looking at what it means for humanity, she sees tremendous potential even if the specific changes are not yet clear to her. An earlier of her books, *All About Flying* (1915, 1919) makes clear that, although humans had wished to fly for millennia, they could not because they "have been led into the wrong paths of approach," trying to fly in flimsy contraptions with beating wings that mimic birds' flight, or basing their experiments on "high-sounding nonsense woven out of the brains of aged ecclesiastics immured in cloistered walls" (2-3). She even tells us her beloved balloons provided a distraction from the main direction of aviation. She describes a history of trial and error, of repeated efforts, of a collective dream humans refused to give up. Yet, in many ways she is also a visionary, looking forward with a clarity about the future of flying we might wish she displayed in speaking of human relations. Even in those days of which she writes, before airplanes were hospitable to people or to cargo, before there was any reason to believe so, she knew the enterprise of aviation would "sway the world. . . This much at least is certain, that, in subduing another element to himself, man has accomplished a feat as great, and with as terrific consequences, as when he pushed his first rude raft into the waves in prehistoric days, or set his earliest sail to penetrate the realms of the great unknown" (121-122).

Thus, Bacon provides a prologue for us, inviting us to know what aviation meant to her generation and, more personally, to experience with her its call to adventure. In contrast to Bacon's brand of aviation, the other women in this study seem to deal with a sophisticated and progressive industry. Without Bacon's story to remind us, however, we lose the knowledge of how far we have come.

Although published only one year later than Bacon's book, Harriet Camac's memoir, *From India to England by Air* (1929), chronicles a completely different generation of air travel. The book consists of her narrative of her journey from Karachi, India, to London, England, a scant six months after Imperial Airways opened the world's first commercial air route between India and England. So new, in fact, was the service that travelers booked passage far more infrequently than did the mail. In a slim, privately published volume of merely 20 pages accompanied by 15 photographs, Camac manages to create a relatively detailed and particularly informative account for consumers of what air travel was like for the commercial traveler in those very early days. In fact, the text and the photographs merge to create a coherent and evocative portrait of Camac's journey.

Far from being mere insertions in the text, Camac uses the photographs to extend her voice, to offer visual corroboration of her words. Ironically, although as photographer she does not appear in any of them, these snapshots reveal as much about the flight as her words and even more about the woman herself. Through them we see with her eyes particular scenes of the journey, and we can use these sights as touchstones to her interpretation of those scenes. Such a standard proves useful, for Camac writes with a personal reticence, a persona objective enough to be almost impersonal. Always the astute observer, cognizant of details that might slip past the careless, Camac steps behind the wording of her text, much as the photographer works behind the camera, in an effort to channel attention solely to her journey. But the reader develops a curiosity about this woman who reports so accurately what appears to us from our perspective a grueling journey. We look for her within the work she has so consciously created.

And so we turn first to the photographs. Of the 15 in the volume, several picture air service accommodations: tents in the midst of vast expanses of flat, barren desert; tired single-story buildings with sparse vegetation planted around foundations and along rock-lined walkways; a refueling station offering no shade in a flat expanse of sand; robe-clad men huddled over clay jars offering the warm water that provides the only suggestion for quenching one's thirst. These images convey a sense of vision flattening bleakness Camac virtually deletes from the text of her memoirs, alluding to it only in her progressively stark descriptions for the desert as her journey continues—the desert that becomes the "unending" desert and, finally, the "interminable" desert before she reaches Alexandria and what she considers the beginnings of civilization. Although Camac refrains from belaboring the point, one senses her growing vexation with the desert conditions and, through the photographs, one understands. Camac also pictures air service officials, the station officers in standard pose side by side looking at the camera with serious faces; work crews laboring under hot sun; and, in what one comes to view as a particularly poignant photo, a tanned and smiling pilot, Capt. A. E. Woodbridge, who within a fortnight of her flight with him burned to death after crashing his plane between Jask and

Lingua. His photograph begins the volume, her obvious tribute to him.

She shows us sights, as well, and not necessarily only the picturesque. In fact, she includes a photograph of oil tanks on the banks of the Tigris and Euphrates. From the air they appear as rows of buttons. And there's the fort at Rokbah which she compares to the fort in *Beau Geste* or the river boat on the Tigris she assures us should be found on Mark Twain's Mississippi River. Most spectacular of all, however, is the arch at Ctesiphon, outside Baghdad, standing as the sentry over magnificent, and isolated, ruins on a flat sandy plain. There, she reminds us, British troops suffered illness and heat during World War I. In many ways they are the images of the vacation traveler, marking the special sights along one's journey.

But one picture tells us much about the mind of Harriet Camac. The foreground captures a camel loaded down with cargo, traveling across the desert as its kin have done for thousands of years. But in the background, poised for destinations as far away in distance as the camel's in time, stands the airplane Camac will soon board to continue her flight. Frozen forever in one frame, both the ageless past and the unknown future face the viewer. Nor was this an accidental pairing, for although coincidence brought them together, Camac photographed them, she tells us, in order to "contrast this ancient means of transport with modern air conveyance" (6). She, too, reminds her reader that she sees the historical impact of her experiences. Through her eyes we see a specific journey turn into an analogy for the twentieth century, a world in which ancient and modern coexist, and more importantly, coexist with our knowledge of them.

Together the text and the photographs demonstrate that Camac knew very well what she was about, on one level depicting for the ordinary person the reality of a journey by air. The text itself contains the sort of detail travelers might need to know, for example that in a closed plane, ordinary clothing may be worn, but one should remember that over the desert the cabin picks up enormous heat and over the Northern regions, such as Switzerland, it becomes extremely cold. In our parlance, the temperature-controlled cabin had not yet been born or conceived of as necessary. She also tells potential travelers of seating arrangements in aircraft—wicker chairs four on each side—and, by implication, suggests the best place to sit: in the front seat because one can stretch one's legs there (a fact of airplane life still true today). She includes descriptions of admittedly primitive airports of her acquaintance, hardly more than landing fields, but describes as well the air service's efforts at lodging and feeding its passengers even in the most exotic of places along its route. And she talks, as well, of the increasing comfort of facilities in Europe. Finally, in the book's closing pages, she assures her reader that the air service, Imperial Airways, runs a very smooth operation, including uniforms in Europe, although such formal clothing would have been unbearable in the heat of desert regions over which she's just flown.

One detail in many guises which Camac uses to emphasize the air service's commitment to its operation recurs in several places throughout her narrative, beginning with the cryptic printing of a quote by Herodotus on the page before Camac's actual text begins, a quote one recognizes as the postal workers' pledge. As the book unfolds, this quotation's connection to every phase of Camac's journey becomes clear. The air service survives because of its ability to deliver mail rapidly over long distances, and the service faces financial penalty if it delays the mail by more than 24 hours of its expected delivery. Thus, the schedules and departure times of its flights revolve around the bundles of mail that need to be transported. Passengers traveling with this fledgling industry are, to all intents and purposes, an afterthought. Thus, Camac finds herself rushing madly across Karachi to board an airplane leaving several hours ahead of schedule and flying long hours in an effort to make up for previous delays in mail delivery caused by monsoon flooding. Indeed, the entire journey flies at the requirements of the mail. Ironically, Camac tells us, such commitment on this particular flight is thwarted by a careless cargo handler in Europe who forgets to load the mail bag onto the plane.

But beyond the official workings of the air service, the reader finds a thorough rendition of the personal implications of air travel in Camac's memoir, for in the chronological accounting of her journey, she wakes the reader at 3:00 AM to catch the only flight of the week, takes the reader in the back of "dilapidated" Ford trucks or taxis to the airfield, and relishes with the reader the warm beer and toast that taste so good to the weary traveler. Through details such as these, she brings her narrative to life for her reader; she encourages her reader to experience the journey with her. Yet even so, she minimizes the inevitable discomforts. For example in telling us that sandwiches packed for lunch had dried out in the 116 F heat of the cabin, she neglects to mention that she and other passengers also rode in that same cabin. Nor does she dwell on the boredom or the unpleasantness of 10 to 12 hour days in the airplane. In many ways, she serves as excellent public relations for air travel, at a time when the industry needed such accounts.

Behind these amenities and anecdotes, Camac also supplies technical information that further defines for the modern reader the state of aviation in the 1920s. She mentions for virtually each leg of her journey the altitude at which they fly, ranging from 500 feet over the desert to 2000 feet later on. Impressed with these altitudes, she nevertheless must remind us that once the flight arrives in Italy, by order of the government forbidding airplane flights over the Alps, they must continue their journey by railway. In an earlier passage explaining a particular method of navigation suited to its location and to airplanes flying without guidance instruments, she describes trenches cut into the desert sand from Baghdad to Gaza to serve as direction markers for pilots, the only direction markers available in the featureless landscape of the area. In another passage, she reports that her flight faced a delay of several hours because a sudden rainstorm soaked the uncovered engines. Ordinarily, according to Camac, open engines cause no problems because the intense heat they generate during flight evaporates any moisture. Standing idle, however, they fill with rainwater and refuse to start for hours.

While these tidbits of information, told in the matter-of-fact tone of one who sees them—quite reasonably—as the state of the industry, amuse the modern reader, Camac's mentions of how her gender affected her trip do not. In the opening pages, she tells us that in Bombay she was told the air service could not handle women passengers and that she would be better off taking the longer ocean route to England. Instead, she traveled by boat during the monsoon season to Karachi, a two days' journey, where she finally obtained air passage. Later, she notes a particularly aggravating complication in which a garbled wireless message announcing the plane's arrival with two passengers left one station manager expecting that the passengers were, of course, male and could share sleeping accommodations. Since no one had even dreamed that a woman might ever be an air passenger, no separate sleeping accommodations existed, and the process of creating alternate arrangements caused a great deal of confusion and chaos.

As an indication of the strength societal conventions have in determining one's perception of appropriate responses, Camac felt compelled to apologize for the trouble *she* was causing everyone! Throughout the journey, Camac faced difficulties that ranged from the mildly amusing to the deeply troublesome. What she confronted was her own essential invisibility to an industry trying to develop a constituency. Again, as did Bacon, she accepts with expressions of only mild annoyance the official attitudes which invalidated any normal use one would expect a woman to have for long-distance flying. Instead, she focuses on the very real improvements in transportation her journey evidences.

In essence, she re-enters the modern world of Europe having traveled over 5000 miles in eight actual days of flying, a journey lasting over three weeks by boat. Her experiences document both the sense of progress one must have felt at the decidedly more efficient method of travel air flight offered and the recognition that it still marked a widely limited and limiting social structure.

Marie Beale's memoir, *The Modern Magic Carpet* (1930), offers an interesting counterpoint to Camac's picture of efficiency, however, for even as she takes us over some of the same territory, literally, that Camac covered, she does so with a different idea in mind. For Camac, the journey was the means by which to get from one place to another more desired place, and the flight's pleasures and adventures were but an extra advantage to its speed; for Beale, the journey itself holds the intrigue. The speed of air travel interests her only in so far as it makes accessible many places previously open only to her imagination. In the tradition of the travel writers of last century, she takes us on a two-week tour of Iraq, Syria, and Palestine in a chartered airplane, operated fittingly enough by the same Imperial Airways we met in Camac's book. However, Beale puts to double use the memoir that takes her to these exotic places, sparked by the imagination we readers find on every page of her volume. Her narrative voice transports us to ancient passageways drawn partly from the pages of history, partly from those of fiction and myth. Her memoir blends the ancient past with the modern age through the controlling metaphor Beale establishes from the very beginning: the airplane is for Beale an enduring magic carpet, and like all good magic carpets, it transports us into realms defined as much by our minds as by physical reality. Beale engages, then, a journey as metaphorical as it is literal, as even her language indicates: "all this realm of history and legend, in which every one's youthful imagination went voyaging, has in after years for the most part remained only the memory of a dream. To see these places and, above all, to fly to them, would be in an astounding way to revive and prolong that dream into waking hours" (1). But, as Beale knew, even dreams have practical sides.

Thus, Beale's account provides new details about the interior of the airplane, and consequently the state of aviation. She mentions the "sanitary packets" of chewing gum which provided the most effective means at the time for combating air sickness. She tells us of the ear stoppers passengers needed to use to keep the drumming of the engines from damaging their ears. She adds the detail that isinglass windows slide open to allow one to adjust temperature. With tact, she divulges that a "dressing room" comprises the rear of the cabin. As the journey progresses, she amuses the modern reader with mentions of the nonchalance and impetuousness that made up her experience with commercial aviation, for example telling of the pilot's willingness to change altitudes at passengers' whim, to allow for a better view, or of her surprise at being able to receive personal messages via the plane's radio, even in such remote places.

As did Camac, Beale sprinkles her text with photographs designed to provide the reader with some idea of her journey. Of open-air markets, mosques, ancient Ninevah, Babylon, and the like, her photographs, however, tell us less about aviation than they do about history. Mostly, they are land-bound shots of attractions listed in ancient histories or described in the holy books of several religions; she does not distinguish between them. Coincidentally, one of her photographs illustrates the contrast between her perspective and Camac's: Unlike Camac's version of the same place, Beale's picture of the Arch of Ctesiphon springs from ground level, closely framed, showing us more than Camac's aerial shot about the vista the arch itself showed to its contemporaries, but less about its geographical place on that vast flat plain. And less, too, about that photograph's place in time, for while Beale's could have been taken anytime from the late 1800s on, Camac's could only have been made during the age of aviation. Different views, yes, but each provides something the other does not; each tells about the woman who included it in her memoir. Beale herself notes that perspectives change from land to air, and if one finds that "all effect of the colored tiled facades was gone" (39) once one ascends, she also finds that "what defects there may be are not seen when one is on the wing" (56). To one unaccustomed, it must seem a form of magic.

Without question, through the images and the written text she includes in the volume, Beale shows a prodigious knowledge of the lands to which she journeys. Aware of cultural achievements long since surpassed and of the layers of grand civilizations indigenous to the Middle East, Beale finds in these lands both a welcome respite from the turmoil of the modern era and a disturbing trend to reject the acquired wisdom of their own heritage. She finds the knowledge of the ancients to be a salvation for her modern age, consequently in her opinion the encroaching desire for westernization of these lands must be seriously misguided. And she mourns it. In a far less Eurocentric vein than we might expect of her, she shows herself to be a lover of museums and a strong supporter of recently renewed efforts to preserve cultural knowledge in museums within the Middle East, commenting in passing that some of those same Western nations that pillaged the region in the past for priceless artifacts should support the local efforts at historical preservation. She immerses herself in the artifacts and anecdotes of the region, telling her reader more about the lands themselves than about the difficulty of getting there.

And this takes us back to that central metaphor, the airplane as magic carpet. It contains the key to Beale's portrait of aviation, and peculiarly, the central focus of her book. With this analogy, Beale ties past and future together: through modern technology, the mysteries of the past can be explored; through that exploration the modern age can find spiritual enrichment. Thus, while air travel seems to shrink into the background of her volume, to be more noteworthy in its absence than in its presence, it underlies everything she has been able to say. Through the speed and flexibility of aviation, she can write of her firsthand encounter with antiquity, for she has traveled more territory, literally and metaphorically, on this journey than she could have by any other means. Through the modern magic of air travel, Beale comes to a more personal understanding of the relationship between present and past. She has, after all, done what used to be the impossible: "Now the modern magic carpet took us back 4,000 years and as many miles in about forty flying hours. Thus comes about a realization of the age-old human longing which inspired the most famous of the Arabian Nights Tales" (72).

Two years later, Marie Beale published another passenger account, this time, as her title tells us, of a *Flight into America's Past: Inca Peaks and Maya Jungles* (1932). Published by G. P. Putnam's Sons, which held a specialized interest in aviation narratives because of G. P. Putnam's marriage to Amelia Earhart, the book again emphasizes the role of aviation in linking present and past, this time in South America. As usual, Beale's focus remains the importance and wonder of rediscovering the magnificence of past civilizations, but also typically as we have come to realize, she credits air travel with making that knowledge possible. In her Preface to the book, she makes the connection explicit, citing the airplane's power to erase distance or isolation and to bring previously inaccessible places within imaginable grasp. She observes that due to the advent of aviation and commercial air routes, ordinary people who for reasons of time or health would never see the wonders of ancient civilizations buried within jungles or deserts could now travel to places previously reserved for explorers or scientists. Indeed, in her eyes the ordinary traveler *becomes* an explorer or a scientist; for her, discovery equates with recovery—the recovery of peoples and ways of life long forgotten.

Thus, this book does not dwell on the wonders of flight, except for Beale's occasional contrast of the world from the air with the world from the ground. That change in perspective of what one perceives still fascinates her. But she wants to instill in her reader a sense of the metaphorical journey she has found so rewarding. Beale even tells us that her purpose is to provide a picture of the far-away, drawn by herself—a confirmed traveler—for that other traveler at home who may long to journey to places beyond the ordinary. The airplane now makes that journey possible, but as integral as it is to making the journey possible, in Beale's mind aviation *per se* takes second place to the destinations it can reach.

Before we leave the passenger narratives, we must examine one more, this one somewhat of an anomaly. Stella Wolfe Murray also devoted much of her life to aviation writing and to a passionate support of women in aviation. One of her pieces of writing, *Woman and Flying* (1929) co-authored by Lady Heath about whom we will hear later, contains a mixture of Murray's essays as well as some by pilots Lady Heath and Lady Bailey, and pilot's widow Mrs. Hinchcliffe. Her preface elucidates the process by which the book took shape: She contacted several women pilots and Capt. W. R. Hinchcliffe's widow for pieces of writing to include; of those she contacted, Lady Bailey referred Murray to previously published articles, Mrs. Hinchcliffe did likewise but also included some additional writing, and Lady Heath responded enthusiastically with a wealth of material, some previously published as well. None of the other contributors to the volume had read Murray's essays, nor necessarily those of the others. The result is a collection of eccentric and highly individualized voices discussing singular accounts of aviation.

Murray, the guiding force and organizer behind the volume, begins it with a disclaimer: While she has been criticized for writing about flying even though she herself is not a pilot, she has gathered the writings of famous women pilots who can relate firsthand experiences for the reader. Furthermore, she goes on to explain her reasons for not learning to fly: age, health, and finances. If the breath of an invalid brushes past us, it is a gentle if persistent and deliberate element of Murray's persona. She returns to oblique references to her health periodically throughout her chapters with a mixture of anger and exhaustion that seems more closely related to attacks of the vapors than we usually read.

Herein lies the problem. Many of Murray's accounts of aviation, and particularly her interpretation of them, give readers reason to question her judgment. In her enthusiasm for the subject she conveys much misinformation, for example, asserting that in America doctors have sent patients on airplane rides involving steep ascents and rapid descents as an "efficient cure" for deafness (104). Other illness can just as surely be cured through aviation, if only the curative powers of air flight garner more exploration, she tells us. "They must have cured something somewhere else in the world by flying, but this is a subject on which I have found it extremely difficult to get data" (104). Murray's proclivity for apparently absurd claims encourages the modern reader to discount much of the information she provides and surely many of her analyses.

But one cannot completely ignore Murray; she has an involvement with aviation that warrants some attention. What we cannot question is the strength of her emotional response to the phenomenon of flight, for it expresses a truth—about her, yes, but also a truth representative of her particular place in history, a place to which Murray was exceptionally sensitive. Her descriptions of air flight focus on its peacefulness, its sense of freedom, and its ability to create a sense of release, especially because she finds it "the only place I know of where the noise of humans is stilled" (106). Behind her descriptions, however, she contrasts her interpretation of flight to her generation's experience of aviation as a medium of war. In fact, she even attributes air sickness to fear of aviation brought about by lingering images the public can't forget of air raids and death-carrying bombers introduced during World War I. She was of the generation most affected by World War I, and as her essays in this volume attest, even in the late 1920s she was still interpreting the world through the events of the war years. She admits to being an enthusiastic militarist in the early days of the war; since then, she has become a committed pacifist. Her passion for women in aviation springs from her persuasion that men created the war, that war perverted the nature of aviation before it could be developed for peaceful uses, and that women because of their inherent maternalism could turn aviation into a boon to peace. Indeed, she seems to have given up on male ability to solve the world's problems and increasingly expresses anger and outrage at the difficulties they place before women who have the inborn abilities and talents to set things straight.

Murray, assuredly more radical than most people in her views of the war, nevertheless expresses a common frame of mind. The war *was* still fresh in the public's collective memory; three of the four passenger narratives we've examined mention the war, although Camac and Beale merely point out the wartime use of various locations, the battles and the troop encampments along their travel route. Other writers during the 1920s, male as well as female, also point to women as the moral hope for worldwide peace, especially in bringing an aura of safety and practicality to the aviation industry. Women's moral superiority and pacifist sentiments were virtually societal givens at the time.

But beyond the somewhat dated reasons for her support of women in flying careers, Murray clearly sees aviation as the wave of the future. The paradigm has shifted, and the future rests in aviation, an industry she firmly believes must open itself to the influence and participation of women. That it seems unwilling to do so causes Murray much consternation. For example, in her rendition of the 1924 ban the International Commission for Air Navigation placed on women pilots, one reads the mixture of anger and frustration in her voice. The anger springs from her sense of injustice at the "sex-prejudice" of the Commission's decision; the frustration from her struggle to explain the nature of the injustice to her reader while still remaining within the standards of language delicacy required by societal usage. Not that Murray wanted to resort to vulgarity; no evidence exists of that. Rather she had to clarify for her reader that women pilots had been banned from obtaining commercial licenses because of their menstrual cycles, which the medical profession—not to mention the conventional wisdom—understood to render women physically and emotionally unstable, hence a danger to any passengers who might be flying with them. Her readers would have been scandalized to see such a reason spelled out in clear words on the printed page, so Murray must use the language of circumlocution, dropping veiled hints about flying at what Lady Heath referred to as "certain periods" (38). As Murray tells us, with very little evidence on which to base a decision, other than a bias against women, "Man, however, naturally picked on woman's main physical disability, as he has always done from time immemorial" (33). No wonder we read so clearly and plainly her frustration.

The anger, too, shows plainly, particularly in relation to two elements of the issue. First this needless ban resulted in actual hardship for one commercially-licensed pilot, Frenchwoman Adrienne Bolland, whose commercial "B" license was summarily invalidated, causing her to lose her air transport company (23-24). Literally put out of business, Bolland, Murray tells us, lost a great deal of money. The aviation establishment, presumably unmoved, did nothing for her. But the treatment of Lady Heath, co-author of Murray's book and clearly an idol for Murray, really caused her disapproval. For Lady Heath, allowing herself to be a test case, was required to demonstrate her flying ability during her menstrual period, which she had to announce to the male pilots who would be evaluating her performance. Today, the outrage rings as loudly s it did to Murray; today, we add to the number of reasons for that outrage.

Murray does not accept with humor, as Bacon did, the pseudo-wisdom of the medical profession, or the International Commission for Air Navigation, or for that matter, men in general. Her tone remains sharper by far than anything we heard from Bacon, or even Camac. In fact, at one point in her volume, Murray feels compelled to offer a kind of apology to the male gender for the harshness of some of her commentary: "Poor dear man! I like him really and hate to be down on him, but truly he has got the world into rather a mess, and it may be that even woman, with all her quality of the new broom, may not be able to help him sweep it clean" (113). But by the time she utters those words, we neither believe nor desire that the apology should be genuine. Yet, Murray troubles us, in ways Bacon did not.

Through it all, inevitably, we wish Murray to have more credibility, to have been able to step beyond personal idiosyncrasies to the objectivity one expects of journalistic writers. Perhaps we want her to step outside her era and to be more modern than she can be. In many ways, Murray more than any of the passenger narrators resembles a young creature barely emerging from the shell: She can sense the new world, but is still encased in the trappings and the fragments of the old. She sees the future shaped by aviation, but as we saw with her analyses, she does not quite comprehend it because she cannot see the future in terms that diverge from the past. Perhaps, in that respect, we ask too much of her.

Unfortunately, the discrepancy between the way Murray wanted to appear—the knowledgeable visionary championing the cause of women in aviation careers—and the way her persona actually translates to readers renders her ultimately an ineffectual voice—except in that almost contagious enthusiasm for flying that grows from the very subjectivity that otherwise plagues her. The magic and the thrill of flying she describes ring true, even today. And just as importantly, the central place she ascribed to aviation in the future world she tried so futilely to see has proven true, as well. For these two observations, we remember Stella Wolfe Murray.

While their narratives treat aviation with varying degrees of thoroughness, these women passengers provide the reader with a view no actual pilot could, for as outsiders they experienced the public face of the industry. More susceptible to the image of aviation than our pilots, they also show us by the issues they address some of the prevalent concerns the public had about fledgling commercial aviation. Their focus on the physical space of the airplane, its relative comfort, and the passage of time during flight reflects the public unfamiliarity with the concept of passenger transportation; their descriptions of visual wonders not seen from ground level reflect the public unfamiliarity with flight in general. The very existence of their books documents a public curiosity about both and perhaps a recognition that technology was on the verge of changing the everyday world in some practical and exciting ways.

Their narratives also identify that from the very beginning, commercial passengers wanted two things from air travel: First, access to new, exotic, and formerly inaccessible places, and second, faster and more convenient schedules for getting there. Today, we take them both for granted.

The British Flyers

While the British women aviators all speak to us with individual voices, they share some commonalities of perspective that tie them together and that also make them as a group very different from the women flyers of the United States. Their common experiences signify a distinctly national identity that underlies all they do for aviation, even as they work to fulfill their own needs and dreams. Such an identity grows naturally from the circumstances that forged British aviation, particularly British circumstances both historical and timely.

For one thing, our aviator/authors do not let us forget that they are subjects of the British crown, and that their position within the British Empire shapes the kind of flying they do, whether they are English women flying over foreign parts of the Empire or European countrywomen of Africa, Australia, or New Zealand flying "back home" to England. In this twilight of the Empire, British aviators of both genders saw as part of their task the linking together of disparate parts of the Empire, forming a more homogenous and accessible Whole of the various colonial holdings. Such a task seemed, in the 1920s and early 1930s, not only possible, but also valuable as a means of enlarging the traffic of resources from the colonies into England and, particularly in the late 1930s, encouraging the availablity of British culture in the farthest reaches of its holdings.

None of these aviators/authors questioned the value of the Empire or its assumptions. In their view—which was, by the way, the prevailing view of much of Western civilization—the Empire, decidedly and rightly Eurocentric, brought a civilizing hand and an improved prosperity to its dominions. This perspective colored aviators' response to the worlds they visited, in the late 1920s and early 1930s almost always flying from one colonial outpost to another and later flying along established colonial commercial air routes. Rarely do we see them mingling by choice with the people indigenous to the areas over which they travel. More often than not, their acquaintance with distant lands channels through the colonial governmental or military representatives of the area. Because of this, what strikes us first today as we read their accounts is that they seem to be always speaking as the perpetual outsider on these journeys. They see the Eastern or African or South American world with the overlay of British custom, British assumptions, British judgments with the baggage of an assumed European superiority. In this, they are products of their time, expressing not a conscious racism but a visceral inability to see beyond a societally limited perspective. The Empire is their inheritance. The British crown their patriotic symbol. The linking of the Empire through efficient and safe aviation their official goal.

Their orientation toward the Empire shaped them in other, more subsidiary, ways. For example, London was the heart of the Empire, even for Jean Batten way out in New Zealand or Lady Bailey in South Africa. Sooner or later all of our aviator/authors found London, and all but Violette De Sibour became members of the London Aeroplane Club, headquartered at Stag Lane Aerodrome. They knew each other, some more closely than others, but they talked aviation matters and learned from each other. Throughout their careers, this professional and personal friendship followed them, the older flyers helping the younger, the younger valuing the experience of the older. In various places within their narratives, these women acknowledge the camaraderie that marked their careers. Furthermore, as they planned their long-distance flights, these aviators dealt with Imperial Airways, in the 1930s just beginning to establish its commercial routes and, at the time, holding an effective monopoly on long-distance commercial flying in the more exotic parts of the Empire. The maps, mechanics, facilities, and airfields, later full aerodromes, of this company served the women aviators essentially as fuel drops, as the dots they connected with their long days of flying over desolate and inaccessible terrain.

That need to fly over wilderness on long, difficult, dangerous, and many times record-breaking journeys also tied these women together. With the exception of Pauline Gower and Dorothy Spicer who set records of their own in other aviation contexts, all these aviator/authors dared the odds with virtually impossible flights. All of them succeeded and felt a companion compulsion to write of their adventures. All of these women knew they had stories to tell, that they had accomplished extraordinary feats, and that other less daring people wanted to read of their lives. From experience, they knew something else, as well.

The women recognized a compelling need to show that long-distance aviation was possible, even advantageous, for women, either as flyers or as passengers, for they rightly understood that only by so doing could aviation fulfill its promise for all people. In virtually all of the narratives from these women, then, we find descriptions of their receptions as women in the various aviation aerodromes, hangars, outstations, and government legations along their routes. Most of them,

Jean Batten and Violette De Sibour are notable exceptions, express varying degrees of frustration and anger at the mean-spiritedness and vagaries of the male aviation establishment.

As they flew, these aviators fulfilled a dream soon to be outdated and superseded by the more general goal of a global network of air communication. For this, their contributions still impress us with their daring and accomplishment.

2

Lady Heath and Lady Bailey: A Marriage of Coincidence

Ironically, coincidence conspired in the 1920s to steadfastly link two women aviators in the public mind; neither found the bonding amusing, for each was a fiercely independent woman unaccustomed to sharing her fame and wanting full credit for her very real accomplishments. Nevertheless, Lady Sophie (Mary) Heath and Lady Mary Bailey shared the spotlight for most of their aviation endeavors, and the press, with some justification, highlighted the rivalry they were supposed to have felt for each other. The circumstances that bound them together are too strong to ignore: both were considered experts by their contemporaries on all matters dealing with women and airplanes; both planned and executed long distance flights of great news value, participated in international air competitions, won awards, and wrote articles for the popular press describing their adventures; at almost exactly the same time, both flew solo over the partially uncharted African continent, covering the distance from Cape Town, South Africa, to Croydon, England, although they did so in opposite directions; and both married wealthy aristocrats who indulged their expensive forays into aviation, but who were content to stay out of the way. We even learn about these two aviators from their chapters in the book Lady Heath co-authored with Stella Wolfe Murray, *Woman and Flying* (1929), of which we've already read, for that book contains their first-person accounts of that famous Cape Town/Croydon circuit, and, in juxtaposing those two narratives, solidlfies the connection between the two women's aviation careers. But these seemingly compelling similarities rest at surface level. More substantively the two women possessed very different personalities that shaped the way they told their stories, as well as the way they perceived their adventures. Their accounts, while on the surface so similar, reflect two very different mindsets.

The Lady Heath that emerges from the printed page can only be described as a formidable woman, headstrong, opinionated, flamboyant. Clearly she loves the spotlight and revels in the telling of her tale, filling it with entertainingly fluid anecdotes and her piquant observations and judgments about just about everything she encounters. Sometimes her pronouncements, for that is the tone she consistently adopts, contradict each other; sometimes they offer sly—even sarcastic—perspectives on people and events; almost always they reveal more about the woman behind the writing than she gives evidence of realizing. But they also provide an insight into the kind of world with which she was familiar, a world in which the kind of flying Lady Heath performed was still a daring feat to be acknowledged and women were still considered incapable of much useful activity outside the home.

Particularly galling to her was the official attitude that, because of the particular dangers they faced as women on long distance flights, women aviators needed special protection, usually translated to mean the denial of permission to fly over certain territories or the requirement that they be accompanied by a male aviator. On her famous flight, Lady Heath had to deal with both contingencies. She recounts that the government of the Sudan refused to permit her to fly over its territory unaccompanied by a man, "owing," she tells us, "to outbreaks among the natives, who killed a district Commissioner in December 1927, and had to be bombed into submission in January" (160). Her flight took place in March, 1928. Under the circumstances, the Sudanese government deemed a male pilot flying an escort plane to be a necessity.

The male pilot chosen to escort her was Capt. R. Bentley, who had already distinguished himself as the first person to fly solo from England to Cape Town. Unfortunately, Lady Heath and he had already flown part of the way together, which resulted in a rather mixed attitude on her part: "I was delighted to hear they [Bentley and his wife] were coming, although of course it took away from the credit of a solo flight" (144). As is fitting for a British Lady, Lady Heath expresses the proper appreciation for his offer to join her on what was potentially a highly dangerous flight; courtesy (and later hindsight that focused on the poor communication system over much of colonial Africa) required that she accept his offer of gallantry. However, there *was* that matter of solo credit that would now be denied her.

To make matters worse, the Bentleys and Lady Heath had very different traveling styles, and the pages of her narrative ripple with undercurrents of snide references to the frustrations and delays the Bentleys caused her, especially Dorys Bentley with whom Lady Heath apparently had little in common and whom she refers to often as "poor Dorys," even as she apparently writes in supposedly complementary or solicitous terms of them: "We did the work on our machines early, and they kindly arranged tennis for me that afternoon. Poor Dorys Bentley looked a bit fagged" (149). She also records the times the Bentleys arrived late at the airfield and the times Dorys felt unwell, such as with "a touch of the sun and a bad tummy" (153), and so delayed the journey even further. Added to their general lack of congeniality, Lady Heath recounts an earlier episode in which she purchased the only set of aviation maps available for Africa from Capt. Bentley, who wanted them returned as he planned his own flight and then promptly lost them, causing delays and problems for Lady Heath. Thus, she must have welcomed the requirement that he accompany her as if it were the plague. As soon as she reached Khartoum, with no comments about the Bentleys at all, "at an impossibly early hour," she tells us, she arrived at the aerodrome and "I pushed off alone" (166). Apparently she had tolerated the Bentleys as long as she was willing to.

That same kind of disingenuous tone marks her comments about Lady Bailey, who really did seem to be perceived by Lady Heath as a rival of sorts. To Lady Bailey went the official credit for the first woman's solo flight across the length of Africa, even though her arrival at Cape Town overlapped the early stages of Lady Heath's flight. Admitting that she had originally planned the trip as a record attempt, so much so that "I did not confide my ambition to fly the continent to anybody, not even to the makers of the machine" for fear someone would try to beat her to it, Lady Heath remarks on Lady Bailey's gallantry, but in somewhat suspect terms:

> As it was, a fortnight after I started from Johannesburg, my gallant little friend Lady Bailey started from England to fly to the Cape on a *Moth* loaned her by the De Haviland Company, and without any real knowledge of her engine or of African conditions. She magnificently accomplished her flight, although she crashed at Tabora and had to obtain a new machine. She was thus the first woman to cover the distance between London and Cape Town flying solo (122).

Far from being a compliment, Lady Heath's attention to Lady Bailey's lack of knowledge of her plane and her crash at Tabora, along with the diminutive "gallant little friend" she uses to describe Lady Bailey, underscore the barely concealed irritation she felt at the thwarting of her own plans for record and fame. Under a very thin veneer of civility, Lady Heath reveals her opinion of Lady Bailey as an undeserving, if lucky, rival.

Unfortunately, Lady Heath is not always astute in the criteria she uses to judge people, and as will several of our aristocratic British aviators, she exudes over the charm and the skill of Italian dictator Benito Mussolini. Her reasons, as did theirs, lack the political edge and humanitarian insight we cannot help but use today in our assessments of his years of dictatorship, but in the late 1920s, Lady Heath was far more concerned with his command over the amenities of colonial society and his ability to provide aid for her. She often needed official aid, it seems, for she frequently recounts problems with obtaining clearance for the stages of her flight, often remarking that Lady Bailey had had no apparent trouble obtaining the permissions and releases she needed for *her* journey.

On one such occasion, Lady Heath's airplane was locked in a hangar outside of Cairo and permission to fly from northern Africa to Europe, unless escorted, was not forthcoming, "although my gallant little friend, Lady Bailey, had been allowed to fly along the northern coast of Africa alone; and, indeed, her approach had been heralded all along the line by letters from the Colonial Office" (172). Announcing to her readers that "I was not allowed to proceed because they did not think I could find my way across the Mediterranean from the African side" and that the British Air Ministry declined to support her request for help in straightening out the mess, Lady Heath tells of the cocktail party in Cairo during which she impulsively decides to cable Mussolini, whom she had never met, for assistance. His cabled reply, "'Have put a seaplane at your disposal'" (172), both flattered and placated Lady Heath. Later, when she arrived in Italy, and met the man, he charmed her further: "Mussolini, that great man who is more of a national monument than an individual, was gracious enough to send for me, and I was greatly struck by his intimate knowledge of details of flying matters. He seemed interested to hear my experiences, and glad of what I was conscientiously able to say about Italian hospitality and efficiency" (173). Clearly, her experience with Mussolini demonstrated his social acumen, especially with aristocratic and famous British Ladies. Perhaps it had a cumulative effect, for throughout her narrative, she consistently praises the attention and treatment she receives from the Italian colonial representatives along her route, while the British seem to be less willing to accommodate her. Incidentally, one cannot help but surmise that her cable from Mussolini, which she carried with her, might have affected her reception at those Italian outposts.

Nevertheless, the Lady Heath who emerges from her narrative is a confirmed Imperialist. Her attraction to Mussolini, aside from his personal charm, springs from the organization she sees as she travels through Italian colonies in Africa. She comments on the communications linkages between Italian outposts and the relative ease with which she finds skilled mechanics at those official airfields. Nor does she disregard the attention she receives as a "guest" of the facilities. She finds herself the center of entertaining social evenings, clearly in her mind the treatment her position and her own charm allow her. The Italians, she seems to be saying, know how to run a colony; the British should learn from them.

Lady Heath fumes often at the waste she attributes to British failure to capitalize on its colonial holdings, especially its failure to establish a communications network throughout its territories. She has had dealings with the system as it exists, as the following exchange describing her difficulties with communication in Kenya illustrates:

> Having so far spent £10 on telegrams to tell the authorities where we were and what we were trying to do for aviation and the British Empire, it was a little trying to find that

> two-thirds of the telegrams were never delivered, and when they did get through were so mutilated as to make no sense. A licensed peep into an official file disclosed something like this: "Owing to her slackness in letting us know her movements we are unable to keep pace with the vagaries of Lady Heath" (156).

She continues by chastising the British colonial system for its slackness in outfitting a proper wireless system within its jurisdiction: "We British are supposed to be the best colonists in the world and yet we permit this terrible state of affairs to go on. . . . In my opinion it is a criminal shame to waste the gallant efforts of [pioneers and explorers] by the carelessness and inefficiency of the postal authorities in one of the richest territories of the Empire" (157). Her concerns seem to be for the safety of aviators and travelers and for the transport of resources from the colonies to Mother England. She sees Africa as a vast and rich continent waiting to be adapted to England's needs.

Between the lines, however, we can hear the response of colonial government officials who have no idea how to deal with an irate Lady Heath. *Their* version of the story, and their assessment of the problems, one may assume, would present a completely different view. The passage above with its mention of "vagaries" hints at the difficulties involved. Indeed, Lady Heath herself tells of her erratic travel plans, her frequent whimsical changes of course, her love of parties and frivolity when she arrives. Her accounts of those arrivals suggest that she envelopes even the most military of outposts by the sheer force of her energy and personality. No wonder she caused official headaches everywhere she went.

Even so, Lady Heath almost eludes us. Her personality, at least the one handed down to us in her pages, remains complex. Equally as obvious as that officious imperious woman we have already met, she can also display a genuinely appealing humility. For example, she refuses to accept credit for accomplishments she feels undeserved. Thus, she remains, if not exactly humble, at least unpretentious when describing a virtually miraculous landing in the African veldt after she succumbed to sunstroke: "The machine had obviously landed itself, and I suppose I must have had some little bit of semi-consciousness when she was coming down, because she was headed into the wind, and had actually not hit any of the trees and thorn scrub which dotted the veldt: but of the landing I remember nothing whatever" (136). Make no mistake, she wants recognition for what she does, such as that heading into the wind that probably saved the airplane from disaster, but Lady Heath refuses credit for all *conscious* efforts involved.

Nor does she have much tolerance for the public adoration of flyers, *per se*. She admits that flying is not all that taxing, that in fact it really is "absurdly easy" as, she assures us, all pilots know. But she acknowledges the "halo of mystery" that surrounds it, primarily because of its inaccessibility for most people (25). From her experience both in flying and servicing her own airplane, however, she finds nothing mysterious about the whole enterprise, although some of it she finds monotonous. On occasion she admits to reading novels in an effort to stave off boredom as her plane flies over the African desert. She strengthens the impression she gives of aviation's simplicity by the luggage she takes with her, the evening gowns and the sports clothes, the tennis rackets and the shoes, hardly the equipment one would expect to take on a dangerous journey.

Clearly, she portrays aviation as part of the paraphernalia of modern life. But the kind of flying Lady Heath performed *was* still dangerous, and to forget that is to undercut her very real courage and daring. As she knew, and as we readers of accounts from all these early aviators come to realize, in case of emergency, help was all too far away and hope of rescue too often remotely slim. Behind all the provocative flamboyance and the theatrical pronouncements that give flavor to her persona, Lady Heath held a perspective on aviation that speaks to these elements of danger and explains in some ways aviation's appeal for her. She addresses the realities of her chosen courses of action in the following terms, that perhaps show us readers a motif in her narrative:

> No nation can advance unless the old ideals of exploration and adventure are lived. There must be lives lost in flying, as in every other step of progress, and as many lives have been lost in the past, but there is no need to run foolish risks. The search for adventure need not entail foolhardiness. Fear is a tonic and danger should be something of a stimulant (138-139).

These words show a serious side to Lady Heath and in some ways undercut the bluster of some of her stories. They show a woman aware of the dangers, maybe even craving them, but certainly not with the capriciousness and impulsiveness we have seen elsewhere in her narrative. These words sound more like those we would expect from a serious professional aviator. Lady Heath has shown us many faces, many paradoxes. One suspects all these contradictions really did coexist in the person who was Lady Heath, just as they most certainly do in the persona through which we know her.

In contrast to that turbulent persona, Lady Mary Bailey seems remarkably taciturn, at least in her accounts of her flight from Croydon to Cape Town and back again. From the very beginning of her narrative, she admits to a certain lack of ease with telling her story. Asked to describe her flight, she tells us she feels "it is difficult to know where to begin or how to describe the wonders that such a trip can reveal" (204). Indeed, her descriptions often repeat that undercurrent, for while she obviously sees the beauty and the variety around her, she does not provide the detail or the explanations that would allow her reader to share those experiences. And rarely does she share an anecdote, although when she does, she displays an understated gift at storytelling. For example, one of her few stories deals with her landing near a village to ask direction and seek repairs. In return for the hospitality of the villagers, she brings white cloth which she gives to the chief when she returns with mechanics. He returns the favor by giving her six eggs and two live chickens, which courtesy demands she accept. Having no other choice she packs the chickens in her luggage compartment for their air journey. "I quite expected to find the chickens very sorry for themselves when I reached Mopti, 100 miles on, but they were perfectly happy and showed a marked disposition to stay where they were among the maps, luggage, and other odds and ends" (215). At the risk of making an outrageous pun, we must consider Lady Bailey unflappable in both experience and the telling of it.

Because Lady Bailey's selection is much shorter than Lady Heath's we have far less time to observe her persona, even though we notice at once its divergence from Lady Heath's, but here fortune smiles kindly on us. On her journey, Lady Bailey also landed in Cairo, she also dealt with the Sudanese government, and she also knew Capt. Bentley. More importantly, she also writes of those experiences, giving us as readers a rare opportunity to find two writers describing virtually the same set of circumstances at virtually the same time. The contrast highlights the substance of their approaches to their adventures.

As did Lady Heath, Lady Bailey finds her plane locked up outside of Cairo and official permission to continue her journey over Sudanese territory denied. Her written reaction shows no emotion, no response on her part, to what she clearly accepts as an official edict she must obey. She does not question the basis of the Sudanese decision, even though she clearly was aware of it: "I was flying alone" (205). Nor does she express dissonance over the decision, arranged she tells us by the Johannesburg *Star*, for Capt. Bentley to accompany her. Indeed, her reaction is quite the opposite, quite the acceptable expression of socially correct response: "I had much appreciated Capt. Bentley's kindness in coming back over this piece of his journey again, and so having enabled me to continue the trip" (206). From her narrative the reader has no indication of what Lady Bailey the flyer thought about her treatment, of whether she found it galling or logical, of whether she wanted to disagree or not. In this respect, she remains a much more private person than Lady Heath.

In order to find a hint at Lady Bailey's visceral response to situations she encounters, we must look at her account of her crash landing at Tabora, from all outside accounts a potentially catastrophic ending to her story, for her narrative provides a glimpse at a complex range of reactions to an extreme event. She tells of getting caught in an air pocket and experiencing a rough landing, which resulted "by the machine coming to rest upside down." With no hesitancy, she takes full blame for the crash, citing her "insufficient care and a lack of knowledge of how to handle a machine in landing at an altitude in the heat of the day" and chastising herself for her foolishness: "I was disgusted with myself" (206). Fortunately, no permanent harm was done to her, and her husband arranged for another plane to be delivered to her in all haste. But Lady Bailey surprises us with her very understated sense of humor, poking fun at herself instead of belaboring her problems: "On the afternoon of the second day after this I watched [Major Meintjes of the South African Air Force] land at Tabora, and comparing my careless way of landing upside down, I realized how it should be done" (206-

7). This quip tells us more about her, however, for it shows a professional aviator learning from her mistakes, driven by an urge to do things correctly.

She's driven also by that pull of adventure that refuses to allow her to take the easy path. Once having completed the journey from Croydon down the east coast of Africa, Lady Bailey finds that it offers her no challenges, save having to again deal with the Sudanese government which she is disinclined to do, so she decides to complete her round trip to Croydon by flying over the largely uncharted west coast of Africa "via the Belgian Congo, French Equatorial Africa, and the Sahara. No one, so far as I could ascertain, had flown that route from South Africa" (211). With no maps and a lack of information about what to expect of terrain, temperatures, weather, and available landing fields, she expects to find the trip arduous, she says, but the trip provides her with a revelation of quite another sort: "I found that in reality I had a chain of aerodromes at my service, and a route provided with many more emergency landing grounds than our own British route, and, as I have said, every expectation and confidence that this was to be one of the great air routes of the world" (212). She attributes the condition of this air route to a joint effort by France and Belgium to link their colonies more directly with Europe. Here the contrast between Lady Heath and Lady Bailey takes on new significance, for while Lady Heath judges her experiences from a subjective and intensely personal point of view, Lady Bailey evaluates hers from a more objective and analytical one. She sees the larger implications of her observations.

Believing she is participating in the beginnings of a pioneer movement, she knows a list of existing airfields along the western route would be useful to later flyers, and she provides it, including distances and approximate flying times between them, as well as a few hints on what clothing and equipment will make flights more comfortable and convenient. For Lady Bailey immediately understands the impact of aviation on the development of the African colonies. Not only will aviation move commerce along, but it will also bring colonists "into contact with civilization," a concern especially of women colonists who feel cut off from medical aid and "isolated from the rest of the world" (215). She, too, sees Africa lying in wait to relieve the problems of Europe, to provide its necessary resources, and to ease its press of population. She, too, ultimately expresses a confirmed Imperialism, if focused more pragmatically than Lady Heath's, at least no less Eurocentric:

> It may be merely the impatience of a woman, but is it not time we ceased to quibble over the exact amount in pounds, shillings, and pence each unit is to contribute to the cost of an All-Red route and looked at the broader Imperial aspect? Trade, they used to say, followed the Flag. To-day and in the future it will also follow the aerodrome, for without speedy communications commercial competition is impossible (217).

Lady Bailey saw realistically the effect of aviation on developing nations, and while that vision was cast in the shape of a world rapidly dissolving, she captured an essential promise of technology in the future. Her vision grew from her own daring and her acceptance of challenge no one expected of women.

Together Lady Heath and Lady Bailey offer a multidimensional image of early British aviation. If they would choose to be seen separately, and safely we can expect they both would, they must face the dominating force of coincidence. When they are remembered, they stand as a pair, for just as their professional lives were interwoven, so are their narratives.

3

Violette De Sibour's "Trip into the Blue"

Safari," Violette De Sibour blithely tells us in *Flying Gypsies* (1930), has no literal meaning in English, although we use it as if it does. It was the name she and her husband Jacques, whom she always refers to as "Jack," chose for the aircraft in which they planned to fly around the world. Safari, translated loosely by De Sibour as "a trip into the blue" (10), describes well the voyage on which she with Jack embarked that long ago September 14, 1928.

Of who they are and what their circumstances happen to be, De Sibour tells us precious little, for such personal details have little to do with their journey, except in the most tangential way. But what she does tell leaves an enigmatic trace for us to follow. From her we learn she and Jack are the parents of "a small son," but she mentions him only three times in the book, once to say they had sent him a wicked-looking dagger for a present. He, who remains nameless in the book, lives in Paris, while they appear to be ensconced in England. Of the nature of this arrangement or of their parental relationship to the child, she makes no mention. De Sibour gives us other intriguing fragments of information, for example mentioning in passing that previously she and Jack lived in the Kenya Colony, where they were friends with famous pilot John Carberry, but of their roles or lives there, she tells us nothing. Of their current circumstances, she admits they jaunt relatively frequently between England, France, Germany, and other European destinations, but again does not reveal whether these trips are pleasure or business. Indeed, we merely speculate that the De Sibours are wealthy, for she tells us that of money for their around-the-world flight they have no worry (5), and as the trip progresses this financial nonchalance seems justified. Furthermore, their extended absence from their lives in England and Europe seems to cause few problems. If we are curious to know more about the people of whom we are reading, we must turn to the 1928 news accounts documenting their departure and journey to fill in the missing bits of information. There we discover that Jack, indeed, has a family pedigree and a title and that Violette is the daughter of a wealthy American merchant in London. By the way, the newspapers also mention that their son, Jean Jacques, is six years old and already loves to fly with his father.

Jack, a former World War I flying ace, was born in France but raised in England. While De Sibour tells us nothing more of his experiences in the war, we learn from the New York *Times* that he was seriously wounded in the war and was hospitalized for an extensive period of time. Nevertheless, as chief pilot for the journey, he takes care of the necessary, if alternately mundane and frustrating, affairs of the airplane itself. He also, we learn as the narrative months pass, tends to become a bit morose, but again De Sibour neglects to elaborate. He is a sportsman, especially a big game hunter, and his search for wilder hunting grounds provides the nominal impetus for the flight, originally planned as a hunting trip to Indochina.

Of De Sibour herself, we learn a bit more. She tells us directly that she is an American, born in Chicago but also raised in England. Of her credentials as the official co-pilot, she admits, "I wasn't much good" (24). If Jack appears infrequently in her narrative, she remains a steady presence, and through her own voice we come to know her. For in addition to narrating the progress of their journey, she also takes us inside her own emotional responses to the journey. She portrays herself as the daring, impetuous member of the partnership. Indeed, it is her suggestion that, if they wish to hunt big game in Indochina, they simply should "fly out." Even as she counters the laughter her idea inspires (for after all, "We had had our own aeroplane for months and flying had become a part of our daily existence"), she, too, labels her idea a "mad suggestion" (4). The picture may not be entirely accurate, however, for as spontaneous as the decision for the flight seems to have been, it preceded an extensive period of planning, mapping, checking weather patterns, etc. The journey may have been, by nature, dangerous, but it appears not to have been foolhardy. Nevertheless, De Sibour seems to glory in being considered madcap, in acting on impulse to make things happen, and in being the one to rush headfirst into adventure, regardless of its nature, and we see her fulfill this role consistently throughout her narrative. Indeed, she characterizes herself as the leader, with Jack appearing as a willing, generally enthusiastic, and usually more practical follower. In fact, she gives voice to this interpretation late in the journey when, after a particularly trying delay that could result in their having to abandon the flight, she observes, "Jack never did take anything seriously. That is part of his charm. I notice that whenever there is any difficult decision to be made, the responsibility is put on me" (164).

Yet, even though the two De Sibours travel together and, as the journey takes on its identity as a social season of sorts, draw other people into their odyssey, the story, nevertheless, remains virtually completely Violette's alone. Jack and the others serve merely as punctuation marks to her adventures—and adventures they are, for from its beginning, scarcely three hours into the voyage that would eventually take ten months, the trip disintegrated into a series of problems and near disasters De Sibour describes with enchanting candor and a touch of wry humor. So hounded was the trip by misfortune, it becomes an object lesson in why such around-the-world flights were so newsworthy in the late 1920s, in that window of time when technology had produced aircraft that *could* travel so far, but commerce and industry had not yet produced a world in which such trips were practical. Far from naive, De Sibour knows the publicity value of the flight, for while travels and air trips were becoming the "in" thing to do for people of privilege, flights to Indochina, especially for a woman, remained exotic and uncommon. Among the gear she insisted on taking, she includes not only a camera, but a motion picture camera, shipping "extra films and cinema spools to various spots on our route" (7-8). Clearly, she planned a memoir of this journey.

She takes with her as well her considerable skill as a storyteller, for she weaves for us a varied and textured account that, as she remarks, "did not lose in the telling" (39). De Sibour is adept at using the pacing of her narrative to maintain the reader's attention, making use of a number of fiction techniques to channel her reader's reactions to the events she recounts. For example, in the midst of recounting a particularly desperate situation concerning landing in Gaya, India, she uses punctuation to carry tone of voice, exclamation points to highlight epiphanies of great importance, question marks to suggest uncertainty, interjections to express emotions too complex for easy discussion:

> We would have plenty of time if we left just after four. But we shaved it too close! . . .Would we make It? . . . Then, in the distance, the lights of Gaya! . . .The victory was not ours yet, as the aerodrome was still to be found. . . The sand—thank God—held solid while we bumped to a standstill. It was not a moment too soon. As quickly as one would switch out a light did night close in on us (206-207).

Just as comfortably, she builds anticipation through frequent use of foreshadow, beginning by characterizing the entire trip from the perspective of the narrator who knows what happened for the reader who does not: "We had wanted to make a dash out of Europe, and then continue in a leisurely fashion, but what we planned and what actually happened seemed to have no connection" (13). Sometimes her foreshadowing takes the form of vague references to uncomfortable feelings and intuitions; sometimes she chooses a more dramatic phrasing: "We all know the old saying about coming events casting shadows. Our next day was to be a long one" (124). In any event, her use of foreshadow invariably ushers in a tale as full of emotion and action as any the reader could wish.

As an illustration, De Sibour's story of a burglar attack during their stay in Shaibah demonstrates her ability to mingle emotions and actions. After warning the reader that "the Shaibah adventures were not over yet," De Sibour paints a picture of a peaceful night's repose interrupted by the "crouching outline of an Arab cautiously entering the room." At this point she adds the information that a neighbor had been killed just outside their door several nights before. With the reader's sense of potential danger at its height, "this figure paused, and then as Jack sprang out of bed, it fled into the darkness, slamming the door as it went." For some, this might be the end of the story, but De Sibour carries it farther, turning her attention to

Jack, who has lost the cord to his pajamas, making pursuit impossible, or at least embarrassing. De Sibour records her own response—giggling—and reassures Jack that perhaps the burglar will return. "Reluctantly, the hero went back to bed. But the burglar never did return" (158-59). With consummate skill, De Sibour leads the reader through several emotions in a relatively few words, leaving a sense of complete narrative within an abbreviated anecdotal form. In effect, she wrings from the story all that it can deliver, for in addition to the emotions aroused within the reader, De Sibour establishes her own response to danger. Her nonchalance in the face of danger, her deft expression of humor when distress would be just as appropriate, if not more so, become her trademarks.

Even so, one of her strengths as storyteller is De Sibour's habit of scattering interludes of description and reverie among her accounts of adventure. The book is stronger for not being merely a collection of anecdotes or a travelogue of exotic sights, but a mixture of impressions filtered through her spirited intelligence. Through her eyes we see an emerging landscape at once both alive and desolate. The starkness startles her. Over and over she describes "sinister coastlines," never ending deserts, and "grim mountains" over which they flew, infusing the alien terrain with human personality. At times she waxes poetic, comparing the Jodhpur taking shape beneath their airplane to "A Grimm's Fairy Tale picture. . .A fortress castle [which] crowned a steep hilltop and dominated the town which lay sleepily stretched at its base" (182). Her descriptions fill the eye and the ear with evocative glimpses from the airplane's cabin window. What she sees frequently stirs the dark corners of the imagination.

Yet she also finds what sounds for all the world to us like beauty: "It was fascinating to watch the country change under one's eyes. Bits of jungled hills came into view. . . .Deep blue and placid stretched the bay of Bengal on our right" (218). Lest we think her opinions have mellowed, however, she follows the above description with, "'Bad—but not as God-all-mighty-awful as I had expected,' was my blasphemous mental verdict on the country" (219). What she longs for and ultimately appreciates resembles very closely an English village dropped into the middle of foreign lands. By the way, she too flies over the Arch of Ctesiphon, but unlike Camac and Beale, mentions it only in passing. Its grandeur is lost on her.

At times her descriptions are purely informative, as when she discusses the oil country of the Near East. With what we today can only consider a charming reminder of days gone by, she marvels at the production of an unfathomable "over four and a half million gallons" of oil daily: "I couldn't begin to grasp the magnitude of it all" (144, 145). Nevertheless, she mentions the effect of the oil export business on the living conditions of the area, mentioning a new hospital along with the squash and tennis courts, the formal gardens, the bungalows, and the golf course. But the oil flares, burning constantly to protect the area from the danger of surplus gas, captured her imagination: "The fire roared like a human monster, blown this way and that by the wind. Through my shoes I could feel the scorching ground" (149). It pleases her, as well, that the oil company "pipes off a few feet" of the gas so villagers living in the area can use it for heating and cooking, at no cost.

Rarely, but more seriously for its infrequency, De Sibour steps outside her role as traveler to address issues she finds compelling, issues in which she finds an injustice. One such occurrence, profound in its simplicity, deals with the dangerous conditions under which French mail pilots carry out their jobs. "Even as I write, there are two men of this line waiting for ransom in a particularly unhealthy spot south of Morocco, where the natives have a bad reputation. I pray for their sakes it may come quickly" (22-23). Her concern, the reader knows, is justified. Evidence of the possible fates of these men surround the De Sibours throughout their flight, for they hear story after story of pilots and travelers killed either by airplane crash or criminal violence, they are warned not to travel alone over wild or unsettled regions, they find concerns raised about the special dangers to women who might be kidnapped to harems, they know from their maps how remote some of their passages must be and how unlikely rescue would be in case of accident. On this journey, De Sibour lives with the knowledge that her own life and Jack's hang by a slender thread between security and danger. She discusses with her readers her reactions to that knowledge.

Again and again, she mentions the nervous strain that she lives with while flying. It drains her, leaving her alternately tired, hysterical, worried, aching. In one poignant account of flying over the "devastating desert" beyond Tripoli, she remembers, "I began measuring distances. How many miles would we have to tramp were we to come down here? I kept wondering to myself, and the reply that suggested itself—about a hundred either way—was anything but reassuring. I began to regret that I had taken along some biscuits and a bottle of mineral water. It was just prolonging death" (61). Yet, while De Sibour never forgets the presence of danger, she declines to be morbid; she does not dwell on the possibilities of unpleasant death, for her response to it is far from simple. For De Sibour, danger equals adventure, and she has always loved adventure. On several occasions she mentions that she and Jack sat enthralled for hours as pilots or military men or explorers spun their stories. This air journey is her chance to have those adventures for herself: "Engine failure here too and we were done. . . . a hundred mile trek without food or water wasn't an attractive prospect—but these are the great moments of life and I was fairly trembling with excitement" (224). In fact, danger added a spice to life that she missed otherwise. In describing an unusually uneventful flight into Saigon, she finds an element missing: "Places like Rangoon or Bangkok, reached after tense moments of fear and danger, are the ones that give the true feeling of exhilaration" (260).

On occasion, it seems that she deliberately seeks out that element of danger, sometimes creating a humorous image of herself as delightfully playful clown. For example, in Baghdad at dinner with King Feisal, she playfully makes a face at him by pushing the end of her nose up. Jack, in horror at what she has done, can only await Feisal's response; fortunately the king laughed (108). Earlier, in Cairo, to combat growing boredom she had masqueraded as an Egyptian woman, complete with face veil and nose ring, in order to play a practical joke on the man in charge of the Cairo Shell Oil Company. Later, in thinking of the joke, she begins to worry about the consequences and so invites her prey to lunch to apologize. Again, she is graciously forgiven, although one cannot help but wonder what this man really felt about being played for a fool. Regardless, his courtesy reigned over his response (86-88). This element of De Sibour's personality allowed her to find the humor in what could otherwise be the discomfort of a long, trouble-ridden journey into strange lands and unknown cultures; it proved useful on many occasions.

But some of her escapades border on more serious potentials. In Sirte, Libya, at an Italian military outpost, surrounded by officers who spoke no English or French, she tries to fill a social void by suggesting that after lunch they should take a picture, and a motion picture at that, of the installation.

"[T]here was an ominous silence. What had I done now? Of course, what an idiot! In this war zone it was as much as one's life was worth to take photographs of the fort, especially moving ones. Our being of French nationality made matters still more grave. Did they think we were spies?" (66-67). Again, fortune smiled on her and the incident ended peacefully. While it seems that she sometimes deliberately courted unnecessary disaster, any discussion of De Sibour's attitude toward danger would be incomplete without a look at her heroic side.

For this, we jump to the latter days of the journey, in Chicago, when in testing their airplane for some friends, the De Sibours face catastrophe. As Jack swings the propeller to start the engine prior to boarding the plane, the engine starts unexpectedly. De Sibour puts this in perspective by telling us, "Had it not been for the blocks, she would have taken off alone, over Jack's dead body" (299). Anchored still so it could not move forward, the plane powered by its open engine, begins digging into the runway, her propeller "being ground to bits on the asphalt." Jack sustains a serious burn in an unsuccessful effort to stop the engine. Within seconds, De Sibour tells us, the plane would probably catch fire. "But I didn't stop to think. Jack, I saw, couldn't get round in time so I dashed forward, sheltering my eyes from the flying splinters. Climbing into the front cockpit, I jammed the throttle back: there was a sudden dead silence, another moment and it would have been too late" (299-300). Gone is the playfulness with which we have seen her face other problems; gone also is her tendency to find humor where one would not expect it. De Sibour realizes just how close a brush with tragedy she and Jack have had.

Beyond the merely, if fascinatingly, personal elements of her memoirs, De Sibour also gives a picture of what both the transportation industry and the political world were like, although she does not do so deliberately. Rather, in discussing the progress of their journey, the lands and governments over which they flew and with which they interacted, she reveals the cultural assumptions that underlie the world as she recognized it in the 1920s and 1930s, a world insulated by wealth and European power, a world of privilege. In many ways, it is also a world that deluded itself, that sought to ignore its inexorable contrasts and tried to preserve a Weltanschauung no longer fitting the reality of the twentieth century.

For De Sibour stands on the edge of a world both expanding and shrinking at the same time. As aviation technology began to take tentative steps toward bringing distant lands closer together and connecting formerly unreachable places, real differences in cultural and political imperatives clashed at many points across the globe. De Sibour, in the midst of some of them, remains in her narrative unscathed by the conflicts. She and Jack appear to be completely and unabashedly disinterested except as spectators to unfolding drama. Indeed, De Sibour displays equal indifference to Mussolini as a brutal dictator and Gandhi as a human and civil rights leader, for although she travels under the protection of Mussolini's military government through Italian colonies in the midst of violent uprisings and later inadvertently shares a landing field with "India's Gandhi," they make no philosophical impact on her narrative account.

But of other elements of her world, she speaks with eloquence. State-of-the-art aviation, at least in the parts of the world De Sibour was traveling, offered primitive accommodations, both in aircraft and land facilities. As she attests, even under ideal conditions, flying was an uncomfortable proposition. She recounts being drenched in rainstorms, overheated in hot regions of the globe, frozen at high altitudes, and shaken by turbulent air pockets. As they journey from outpost to outpost, she describes makeshift landing fields made from racetracks located miles from town, "rotten little fields" that make landing a gamble, stark and isolated buildings rising out of nowhere, and an unpredictable availability of even the most modest accommodations. Her depiction of Jask, a centerpiece of the fledgling airline network between India and England, captures the flavor of the airways in existence at the time she wrote: "So this was Jask—a place of ill-omen to flyers. A few dilapidated houses, badly in need of a coat of white-wash, was all there was to be seen. It was the typical telegraph station. . . . this God-forsaken place" (169-170). Yet, as uncomfortable as she found air travel, De Sibour also recognized its advantages over other existing modes of travel and its importance in shaping the future.

Of other modes of travel, De Sibour had prior knowledge. Even on this journey, she and Jack often had to travel by land, rail, or boat to reach their destinations or to continue the trip while the Safari was being repaired. Early in the journey, she expresses her amazement at the speed of air travel, giving the reader concrete details to back her observation: "Once before on a visit to Morocco I had made this same journey by car, taking a good two days in the process. I could hardly believe my eyes when, after an hour and thirty minutes, we sighted Fez" (31). Or again, she notes that they were able to fly 590 miles in a scant two and a half hours (36). Although they suffered frustrating and lengthy delays *on the ground* during their air voyage, De Sibour credits those detainments to the exasperatingly slow delivery of repair parts and the nonchalance of native officials in the isolated locations they visit. Flight itself, as she has evidence to prove, saves enormous amounts of time.

Another advantage she sees for aviation deals more with the aesthetic than the practical. From the air, one gains a completely different perspective than one holds from the ground. It can change one's perceptions, if not the reality. For example, the Suez Canal, "the great waterway of the East," looks "ridiculous from the air, hardly more than a ditch" (90), while Baghdad from the air becomes "all that my imagination craved. A city of the Arabian Nights—sleepy and Eastern looking (99). Unfortunately, "Baghdad, the actual town, turned out to be not such an enchanting spot as I had fancied from the sky" (105). De Sibour frequently describes the land as it passes beneath her, using poetic language to convey a sense of wonder. For example, she talks of Spain forming as they fly over it, as if it were a fluid shape, and later she comments on the sharp distinction between the fertile strip of land bordering the Nile and the desert that adjoins it. Throughout her narrative she offers the reader a glimpse of what she sees from the vantage point of her plane. It seems that along with adventure, De Sibour craves new sensation, new sights, and she drinks them in with appreciation.

Not all of the experience of air travel takes place on landing fields or in the air, however, and De Sibour chronicles a journey met at virtually every corner with hospitality, whether it be from local potentates, military officials, representatives of the Anglo-Persian Oil Company, other aviators, or legations from the reigning British, Italian, or French colonial governments. They enter a social whirl of parties and dining featuring the best the locality could offer, calling for the evening dress and Jack's dinner jacket she carefully included in their necessities for the flight. Indeed, at one point several months into the trip, she comments that she can "count on the fingers of one hand" the occasions on which she and Jack had to dine alone (178). Most of her hosts, especially the government or business officials and the local potentates—including King Feisal, enter her narrative and make speedy exits, for her mention of them seems to be a courtesy she extends to them. A few, however, working aviators and wanderers like herself and Jack, play more important and recurring parts, becoming travel companions for a period of time or acquaintances one meets periodically at various locations throughout the voyage. In scenes reminiscent of Ernest Hemingway's fiesta in Pamplona in *The Sun Also Rises* (1926), she tells of two newly-acquired friends who shared their spur-of-the-moment, party filled, excursion from Baghdad to Tehran, where unfortunately Jack was sidelined from the revelry by appendicitis and a subsequent operation. While the jaunt served as an interlude in their journey, due to long delays in repairing Safari, it becomes a necessary piece of the whole narrative, for with it, and other less dramatic episodes, De Sibour paints a picture of the easy fellowship and nonchalance shared among professional aviators. She and Jack feel honored to travel in their midst. Part of this camaraderie she attributes to the shared sense of danger they all feel, knowing that sometimes traveling in company with another airplane can be the margin of safety. Still, she introduces her readers to a group of fun-loving, highly skilled, hard drinking men for whom the sound of adventure is as much a siren song as it is for her and Jack.

Most of the social life she recounts, however, has its basis in the European network spanning Africa, the Near East, India, and Indochina. The De Sibours travel from one European outpost to another in their trek across the globe, and their companions are the British, French, or Italians they meet along the way. De Sibour returns to the touches of home for comfort, the tea and scones she was served in Karachi, India, or the house of Bertram Thomas that "seemed like an English country house with only an occasional touch of exotic colour to remind one that it was Arabia" (172). Unlike many European travelers of the day, however, they do occasionally mingle with the indigenous population. In fact, their hunting trip in French Indochina involves their living in the jungle and in primitive villages for several days. She marvels at the primitive huts shared by as many as twenty people and pronounces the standard diet of eggs and unseasoned rice unsatisfactory "for any length of time" (278). But most of her contact with the indigenous population is with the potentates, the powerful local figures that she recognizes as merely locally important. Even King Feisal, perhaps the most famous of her local hosts, elicits more humor than respect from her. She knows, in the subconscious level of world view, that under reigning

Imperialism, local power exists at the grace of colonial power, European power: "The natives regarded us with curiosity—and respect when they recognized under whose care we were" (173).

That assumption of unquestioned respect for Europeans shapes much of the De Sibours' itinerary, allowing them to view sightseeing in the midst of a rebel uprising in Italian colonial Northern Africa as a "thrilling adventure" (71). Indeed, in describing their escorted motor trip into some of the disputed territory, she tells us, "I couldn't take it seriously then, but a day later when we found the scar of a bullet on our propeller, I began to realize that this was not a musical comedy. . . . We sat back and pondered that. Jack and I were enjoying life" (72). Even warnings that the government would not take responsibility for their safety had no effect.

Readers in a post-colonial world look at these events from a more objective perspective, resulting—if we may borrow terminology from literary criticism—in our finding De Sibour to be a "naive narrator" in some parts of her narrative. We as readers see events far differently from the way she presents them; we see more of the undercurrents, the subtext, than she. Her account of a village fist fight resulting from a land dispute, gives us perhaps the clearest example of this dual perspective. In her words:

> As we approached the mud hamlet, we could see the inhabitants standing in little sullen groups. Active fighting had ceased, but hostility was still smouldering.
>
> At the sight of Colonel Peake, however, the atmosphere changed. He is both beloved and respected by the natives. Their chief came forward, with an ingratiating smile, bowing us to his house. . .
>
> The fighting was over, and all that remained for the mounted police to do was to take the ringleaders to cool down in the Amman jail. "[The chief] is pretending regret that we should have been bothered with the affairs of his village," wound up Colonel Peake, "and I am pretending to be very angry" (93-94).

Post-colonial readers are more apt to see the local population responding pragmatically to a currently more powerful force than we are to see them reacting with affection and respect. The chief's hospitality becomes a required act of deference, while the charade at the end represents a formulaic dance of domination and submission, with both parties knowing the rules. But in the late 1920s, such conclusions were literally unthinkable for most Europeans, secure as they were in the rightness of their civilization. De Sibour is no exception.

What ultimately, then, is our assessment of Violette De Sibour. One would not make too great a claim to assert that De Sibour's narrative persona proves to be one of her strengths. As we hear her story in her own voice, we gain a sense of knowing this somewhat unpredictable optimist with her strong opinions and her enthusiasm for living. She demonstrates herself to be a woman of great daring who, if she sometimes leans toward the foolhardy, also bends toward the courageous and heroic. The journey she shares with us carries her beyond her own personal fears and discomforts into a world few women, indeed few men, of her upbringing and social class ever experience. At various points in her narrative she acknowledges this, comparing herself and Jack to Adam and Eve or Mr. and Mrs. Gulliver.

But always throughout the narrative, the reader confronts an unrelentingly honest Violette De Sibour who refuses to shy away from telling us what she perceives to be the truth. So she can say to us without shame that flying made her uneasy, she can admit to experimenting with opium in India, she can counsel us that flying is usually uncomfortable, she can recount her frustration and despair at the problems they encountered on their journey. She tells us what she wants us to know; she unabashedly declines to tell us more.

Ultimately, then, Violette De Sibour is a woman who defies stereotypes, the literary ones as well as the societal ones. Through her eyes we gain enormous insight into the realities of a world just beginning to adjust to the twentieth century.

4

Mary Bruce: The Flight of the Bluebird

There is no more subversive act
than the act of writing from a
woman's experience of life using
a woman's judgment.
—Ursula K. LeGuin, 1986

At first glance this epigraph seems inappropriate for a volume of memoirs as apparently conventional as *The Bluebird's Flight* (1931) by the Hon. Mrs. Victor Bruce, as she identifies herself on the title page. On the surface, the book chronicles the journey of a British aristocrat who, thanks to the Dictaphone she decides to include instead of a radio, records for her husband her adventures and her observations as she flies from one colonial outpost (British or French, predominantly) to another in the Middle East and Orient on a flight that even takes her over the United States to find her mother's childhood home in New Albany, Indiana. Throughout the book, she reminds us of her concern for the husband and son she left behind and of her love for England, to which she longs to return. Despite the circumstances of her voyage—she *was*, after all, a woman alone in a tiny single-engine aeroplane on an around-the-world solo flight no one else of either gender had succeeded in finishing—she conveys the image of proud crown subject and faithful mother, loving wife, and dutiful daughter to her family.

Indeed, her own name, Mary, is mentioned only once in the book, and then merely in passing in the latter pages of her story; fittingly, she utters it to herself. What one comes to understand from her memoir is that while she may appear to the world as the Hon. Mrs. Victor Bruce submerged so completely in her husband's identity and status that even her name seems lost, she remains Mary to herself. And the story she writes *belongs* to Mary Bruce. She makes it so through innumerable acts of subversion in the text, subtly drawn needles that puncture the seamless fabric of her oh-so-conventional narrative and leave the reader (the female reader? the astute reader?) with the unmistakable knowledge that Mary Bruce, bound by the double bonds of gender and social standing, has told her own story anyway, even if she has had to masquerade as "Hon. Mrs. Victor" to do it.

The subversion begins early. Colonel the Master Sempill (the name with which he signs his contribution to the book) pens the foreword to the volume. In a voice remarkable for its blandness, he uses his page and a half to usurp Mary Bruce's story, telling the reader that, first of all, he had been responsible for Mrs. Bruce's flying: Knowing her well, he "felt that only in that medium could her insatiable desire for speed be satisfied with safety" (xi). He thought flying would be best for her; he did not, however, expect her to fly around the world on her first voyage, so he admits to being "frankly incredulous." Clearly he finds no sense or reason for her inexplicable haste in attempting an unheard of voyage that few before her had ever tried. We can imagine him shaking his head as one would when dealing with a recalcitrant child, for along with her insatiable desire for speed, this impetuousness marks her in his studied viewpoint as an emotional, unreasonable female.

Faced, however, with the undeniable fact that Mary Bruce survived her voyage quite well, thank you, and that she did, indeed, accomplish things no pilot of either gender had done before, Colonel the Master Sempill can only say that "she does not give adequate credit to her own skill and pluck" (xi), nor does he elaborate on that statement. On the strength of her achievements, Colonel the Master even deigns to admit her into the magic circle of the "world of flying," for he says, "we can consider we are fortunate in having secured her support for the new [aviation] movement" (xii). She clearly has received his stamp of approval. But, there is a price to pay: He says, "we have secured her support," not "she has won our support;" we have given, instead of she has earned. As Mary Bruce knows, gifts can be given to anyone; prizes, however, must be won by those worthy of them. Colonel the Master by taking the active role in his account of their relationship has appropriated Mary Bruce's will and consciousness.

So what is Mary Bruce to do? She cannot insult the friend who now stands as her sponsor; maybe she doesn't even want to, or know that she wants to. But deep inside she must tell another version of the story—her story, her decisions, her victories, her emotions, her journey. Social status and gender dictate the feminine image assigned to her as a respectable woman, so Mary Bruce uses it: Chapter One, entitled "An Inspiration," tells of her impulsive decision to buy an airplane because the dress she tried on didn't fit:

> I had never cared about flying, and in fact had only once been up in the air; although I do a great deal of motor-boat and car racing, I had always been afraid of flying. I used to tell my friends that I should never fly and that sometimes I even hated butterflies, or anything with wings, and that it actually made me dizzy to look at my own foot. That was my outlook so far as flying was concerned until this day when I spied the little machine in that shop window (1-2).

And having seen the plane, she of course decided she wanted to fly around the world in it; why else buy it?

On first reading, this seems the worst kind of reckless frivolity (one is tempted to say flightiness, but refrains), the kind embedded in all those negative stereotypes of femaleness: the emotional instead of rational thought, the lack of planning, the uncontrollable shopping if nothing else. That's the Hon. Mrs. Victor speaking. On second reading this passage appears to be Mrs. Bruce's bow to societal expectations, for it softens the edge of what she tells of in the book. After all, a woman undertaking a dangerous journey on a whim projects a much less threatening image than one who plans and prepares with full knowledge that she is attempting a journey no man has been able to accomplish yet. Women are *supposed* to be irrational creatures of unpredictable emotion, so she gives her readers this image.

On third reading, however, Mary Bruce's words take on a more subversive edge. For here, on the very first two pages of her book, she has taken back her decisions and accomplishments —her own story; she has divorced them from any influence Colonel the Master or Hon. Mr. Victor might have had. Born of her own decision in a situation she defines as being clearly without male interference, Mary Bruce's journey belongs to her. She makes her reasons so outrageously stereotypically female that no male would even try to claim them. They are in her milieu identifiably a woman's reasons, voiced in the only convincing tones available to her. Colonel the Master and Hon. Mr. Victor could recognize that feminine voice, even perhaps expect it and be amused by it. So comforted, lulled, would they be by the tones, they would not need to see the implications. They would not need even to hear Mary Bruce's voice, issuing her own declaration of independence.

How do we know Mary Bruce exists within the inner structures of this volume? How can we be so sure that there exists a Mary Bruce distinguishable from the Hon. Mrs. Victor Bruce who signs the volume? To answer this, we must examine both what is said and the way in which it is said; we must determine the author's awareness of her role as storyteller; we must deal with both the events and the implications of the way they are recorded. Mary Bruce, first of all, can tell a convincing story. She controls the buildup of suspense with skillful timing and selection of events. An example suffices to illustrate; she describes one particularly long, potentially dangerous flight over open sea:

> And now beneath me I could see little white horses on the water. Cold fear came over me. The wind was getting up, and it was against me! It is on these occasions that one begins to hear and fancy all kinds of noises in the engine. I listened. I was certain something was wrong. . . There was not the slightest doubt: the note of the engine had changed (161).

The appropriate exclamation point, the rhythm of the short sentences, the tone all convey impending action. As surely as she has built the suspense, she capitalizes on it: the problem she discovers is merely a leaky exhaust pipe causing no danger. As storyteller, Mary Bruce has taken an incident of no import and created from it a moment of drama, drawing entirely on her own reactions for its texture and impact.

Without question, Mary Bruce knows her role as storyteller; she knows that one can control the impressions one gives by the manner and selection of what one tells. Take for one final example her account of tiger hunting in India. She tells us that she asked to be photographed with the dead tiger, which had been shot by one of the men. Why, we might ask, does she want such a picture? Her reply: "Now, when I tell the story of that tiger, sometimes I say we did it, sometimes I say they did it, but in America, of course, I said *I* did it" (117). Storytellers are in charge. They are the shapers of their narratives, as Mary Bruce so obviously knows. So, to return to the original question of how we can know that Mary Bruce is really telling a much different story from the one the Hon. Mrs. Victor Bruce seems to be recording, the answer is so simple; the woman who lived the adventures and carried out the journey chronicled in the book, who dared danger from hostile people and unfamiliar terrain, who challenged nature and precedent and still won, who felt so vividly the exaltation and the despair recorded in those pages, who shaped and controlled the narrative so skillfully, showed a quality of mind and spirit incompatible with the woman who would fly around the world merely on a whim with no idea of the dangers and challenges ahead. There is more to Mary Bruce than she admits forthrightly.

So, who is Mary Bruce and how does her story differ from Hon. Mrs. Victor Bruce's? Evidence shows us that even many of her friends failed to distinguish between the two, for they heard only the voice of Hon. Mrs. Victor Bruce. They expected her to fail or give up on her voyage, as is cheerfully relayed to her by her husband. On the first evening of her journey, alone in Frankfort after having her wireless burn out and having fought storms and route changes all day before finally landing for fresh weather reports and new maps, she reports to her readers her telephone conversation with her husband, and she concedes to us that he "was overjoyed, however, on learning that I had made so good a journey, and told me none of my friends had expected that I should succeed in getting out of England. They thought I should probably come down before reaching the

Channel" (16). One can almost forgive them these many years later for seeing only the facade; they saw what they expected to see. They saw what society taught them they would see.

Besides, Mary Bruce was used to people thinking her irresponsible or insane for even imagining the flight, much less going about making arrangements for it. She tells us as much when she describes the reactions of flight instructors, engineers and mechanics, mapmakers, etc., as she tells them she plans to leave on her journey within two weeks. In fact, she even deals with the not-too-polite versions of their scorn: "Some interested person on a visit to Brough who knew very little about aeroplanes, seeing the [call] letters [G.-A.B.D.S.] being painted on the wings, tactlessly asked the head engineer, who had been lying under the machine nearly half the night, what they stood for. `A B_____ Daft Stunt' came the tired reply" (9). To be fair, one can empathize with what must have appeared to him and to others she encountered as just another spoiled and arrogant aristocrat insisting on privileged treatment. Hon. Mrs. Victor Bruce, however, seems impervious to their remarks and raised eyebrows; she manages to ignore them.

Mary Bruce cannot; she wants credit as a serious aviator. In response to the astonishment expressed by the people at the Map Department of the Automobile Association who understandably "thought I must be completely mad to be organising a flight round the world before learning to fly," she informs us as her readers, "But there was a reason for this. . . I had been told that the only time of year possible to make the flight would be the autumn, and that unless the maps were ordered immediately they would not be ready in time" (6). As she carefully explains to her readers, with such a rigid timetable Mary Bruce must begin preparing for her flight. She, too, learns to service the aeroplane and to make its repairs. Later on, a broken propeller offers no problems; she can change it with the spare she carries. Dragging the plane across a bumpy clearing in the Indochina jungle to gain a more advantageous take-off position offers no problems, either; she's equal to the tasks of both selecting a favorable position for take-off with all the mathematical calculations, geometry, and physics it requires and physically moving the aeroplane across rough ground by herself. She proves with action that she can succeed through more than merely buying the services of others.

But this is not to say that Mary Bruce openly defies the societal conventions surrounding women of her social class in those days. Indeed, she complies with their expectations—to a point. The facade she takes on requires her to be demur and helpless, to recognize her own weaknesses as a female, not an easy task while she is talking of her extraordinary adventures. She accomplishes it, however, partially through the simple and consistent device of assigning credit for her actions to forces outside herself. Sometimes this means that she credits other people for her survival: She escapes an early crash because, she tells us, "it is thanks only to the wonderful instruction given me in England that I came out of it so well" (27). Or on a later occasion, "I shall always look back upon that day's flight and its safe ending as a miracle wrought in answer to the prayers of the priest at Lakhon and the St. Christopher medal" (126). She credits luck as well, telling us on occasion that luck made her mark her map so carefully (153), or that typically when she found herself in danger "something always turned up at the eleventh hour—some miraculous aid—to straighten things out" (37). With seemingly straightforward humility, she omits any mention of her own skill or courage or initiative. In these passages she refrains from identifying herself or her skill as a possible agent of the miraculous escapes, nor does she mention that all the wonderful training in the world can be of little help unless the one alone in the cockpit maintains the presence of mind to act on it.

So willing is she, in fact, to give credit to others that she even includes the Hon. Mr. Victor in the kudos, for on the moment of embarking on her historical voyage, she can pause, she tells us, to think, "He was the brave one, for how much worse is it to wait for news than to be the item of news" (14); significantly, if one reads carefully, one notices that she couches this nod to her husband in the form of a question instead of a statement. How much worse is it in real terms, Dear Reader, to wait for news than to attempt present danger? Thus, she qualifies her praise with a subtle challenge to reason. And this qualification provides a clue that Mary Bruce chafes under the restrictions society as she knows it places on acceptable behavior for women.

There are other clues. If she gives away credit for her own achievements, she also refuses to accept blame for mistakes or mishaps that are not hers. On one occasion, facing a very real possibility of crashing in shark infested waters, she determines to leave a Dictaphone message to her husband informing him that the crash, which she likely would not survive, was caused by faulty equipment, not pilot error. If she refrains from bragging about her accomplishments and records, she also manages to keep track of them. On many occasions during this flight, Mary Bruce becomes the first person to fly over some body of water—in fact, on crossing the Yellow Sea she set a record for the second longest flight over a body of water, second only to Charles Lindbergh and his Atlantic crossing—or the first Britisher to enter some country or the first woman to fly over certain territory. She mentions these records only in passing, but she records them, nevertheless. Usually she tells us also that the awards held meaning for her. On receiving the Order of the Million Elephants and the White Umbrella for being the first Briton and also the first woman to fly to French Indochina, for example, she records that "The warm greetings came at a time when I much needed sympathy and encouragement, and I felt renewed in spirit as I took off from Hanoi Aerodrome" (130)

In addition, if she credits the men around her for their aid, she also sometimes presents in ridiculous light some of those male precautions foisted on her, a case of male chivalry going awry. In India, for example, Captain Pendlebury, Brigade Major at Allahabad, insists that she must carry a revolver because she will be flying over "some of the worst jungle country in the world" (87). She declines the offer, but he insists; then having forgotten the revolver, he must run home to retrieve it. He arrives back at the airport only moments before take-off and drops the revolver into the cockpit. Twenty minutes later, the cockpit fills with smoke. "For the moment I thought that the wretched thing had gone off, and I nearly jumped out of the machine in fright. Then I saw what had happened" (87). The revolver so hastily dropped into the plane by the Brigade Major had broken a canister of smoke bomb. Mary Bruce was not impressed. Her rendition assures that we as readers aren't impressed by such bumbling gallantry, either.

Perhaps the subtlest example of Mary Bruce's need to record her own achievement rests in her use of juxtaposition, and it requires careful reading to recognize. In describing her experiences in the United States, Mary Bruce compliments rightly the air mail pilots she meets: "They are the heroes one hears nothing about, for they silently carry out their job night and day, running colossal risks in order to get the air mail through on time" (204). As Charles Lindbergh recorded in his book *We* (1927), during the 1920s air mail was in its infancy, and pilots frequently faced danger from adverse weather, poor visibility, infrequent landing fields, and other assorted problems. Antoine de Saint Exupéry writes in *Wind, Sand, and Stars* (1932) of his own experiences as air mail pilot over the Mediterranean, emphasizing also the danger pilots faced daily and the rituals they maintained to give themselves courage. Mary Bruce does not exaggerate in her praise of air mail pilots. Nevertheless, several pages later in her book, on the way to Medford, Oregon, she tells of ice collecting on the windscreen and wings of her plane, of the "terrifying" landscape of twisting gorges and rocky passes she must transverse, and of her ultimate safe arrival at the airport: "Astonishment was on the faces of all the officials. They couldn't understand how I had got through safely, for the air mail had not arrived that day, and was not likely to for some time, by the look of the weather" (218). She succeeded under circumstances that daunted the air mail pilots. Mary Bruce draws no connections in the text of her book; she does not point the reader toward any conclusions. In fact, without emphasizing the issue at all, Mary Bruce establishes her own credentials for bravery and skill under pressure. She creates the written record and stands aside for the reader to find it.

She uses this technique successfully in other circumstances, even more subversive than the one mentioned above. She actually gently ridicules the feminine stereotype which she finds so confining to herself as a skilled and articulate woman; she may give surface validity to the image of the helpless and fragile female in need of the strength and courage of male protectors, but she clearly identifies it to the careful reader as part of the facade she feels she must adopt. Again, to demonstrate Mary Bruce's dual message, we must turn to the juxtaposition of closely related episodes. The first of these shows Hon. Mrs. Victor Bruce using her feminine wiles to manage a potentially dangerous situation. After an emergency landing over restricted air space at a public stadium in Angora, Turkey, which she accomplishes by dropping a smoke bomb to clear out spectators, then taking off her helmet to show by her hair she is a woman and "that they mustn't dare to hurt me," Mrs. Bruce finds herself taken like a prisoner to the Governor's:

> He understood French, and I immediately explained to him

> that I had made this long and arduous journey just for the purpose of saying'How do you do?.'
>
> He immediately smiled and in excellent French said: 'You have come all this way to see me?'
>
> I hung my head and replied,'Yes!'
>
> 'Ah,!' he said; 'mon petit "oiseau bleu"! You have the freedom of Turkey. I will immediately signal to all aerodromes and tell them to give you every assistance possible.'
>
> I smiled to myself when I thought of the warning my friends in England had given me about Turkey and the fate which awaited me if I landed there (25-26).

Clearly playing her role as the coquettish female, Mrs. Bruce wins a battle of wits with the Turkish Governor by outrageous flattery. Her narrative here emphasizes the ploy while at the same time revealing her deliberate and conscious use of it. This Mrs. Bruce knows how to play the game by age-old established rules. Mary Bruce openly admits the ploy to the reader; in fact, she invites us to chuckle with her.

The companion episode, however, offers no such clearly marked irony. Shortly later on the journey, Mary Bruce crashes in the Arabian desert, damaging the aeroplane only slightly, but also mooring it in soft sand from which the plane cannot gain enough momentum to take off. Stranded in the desert, surrounded by "hostile Baluchi tribesmen," Mary Bruce must use every bit of her personal courage and ingenuity to avoid disaster. "I knew it might be dangerous to show fear, so instead I smiled and went up to them, shaking hands with a very fierce-looking man with a long black beard, who appeared to be the chief. This seemed to please them and their scowls changed to grins. To show my complete unconcern, I made friends with the children" (49). During the three days she remains with them, she protects her aeroplane, disagrees with the tribesmen about covering her face because she is a woman, confronts a band of robbers who pillage her plane and sexually threaten her, and finally walks across the desert to a village where she can find help. In short, she manages her survival and her own rescue quite well. Meanwhile, officials at Jask, notified of her disappearance, send a rescue party to her aid. After all the danger is passed, they find Mary Bruce.

Here, again, she adopts a feminine posture, becoming for their benefit the emotional female: "I had been able to keep up my nerve quite well during the past three days, but I had always felt that if ever I heard an English voice, I should break down, for the joy would be too much, and as they came nearer I felt the tears coming into my eyes, hard as I tried to keep them back" (71). But Mary Bruce does not quite disappear during this part of the "rescue," for she still expresses concern for the tribesmen and villagers: "I was afraid that if I showed that I was upset, they might think that the Baluchis had ill-treated me and fire upon them" (71). Mary Bruce cannot completely turn into the helplessly simpering female. The official report notes merely that she is a "woman of great nerve and endurance. Few, if any, would have met these troubles with her spirit" (70); so much for her ingenuity, courage, and survival skills.

The report, however, contains truth along with the truth it omits. Mary Bruce remains a spirited woman throughout her journey, and in the process of completing it, she learns many things, or perhaps realizes many things, that she could not have seen without the journey. Perhaps because of her reliance on her Dictaphone journal, we readers watch her change and grow from the rather provincial aristocrat who begins the journey to a more open, less judgmental woman at the end of it. The change, however, is not complete, for Mrs. Bruce remains inevitably caught between two worlds, and both lay strong claims on her. Suffice it to say, she begins to develop a sense of cultural awareness many never reach, for as she travels she analyzes, sometimes misinterpreting, the lifestyles and cultures around her. And sometimes, just sometimes, she forges new understandings that lift her beyond the imperial attitude one expects, and often finds, in her world view.

After all, she had to overcome tremendous barriers. Even the journey's design itself insulated her from the cultures she passed through. Flying from one colonial outpost to another, she planned to be surrounded by the English, entertained with English teas and dinner parties, hearing the English language and enjoying the fruits of the Empire. To a very great extent, this occurred, but as so often happens, the unexpected caught up with her. Early in her voyage she found herself mingling with other cultures. For example, in Turkey, she met the Governor's mother living in traditional seclusion and came face to face with another culture's opinion of her: "They were both so interested; it was something new in their secluded lives to find a woman flying alone on a journey of this kind, although I am not quite sure that the old lady approved of such feminine independence. I expect, however, she thought it was just part of the new era of women's emancipation which is now prevailing in Turkey" (32). Her naive interpretation of the event, even though she recognizes probable disapproval, still retains its English bias; Mrs. Bruce cannot quite see beyond her own cultural boundaries to place her actions in the perspective of another culture.

Shortly later, after her crash landing in the Arabian desert, she finds herself actually living the life of a woman in another culture, for when the Baluchi tribesmen discover she is a female, they make her tote the water and perform the heavy tasks which fall to women in their culture. Chivalry in English terms has no meaning for them. While she complies from necessity with some of the demands placed on her, even in covering her legs and arms as is the dress code for Baluchi women, she refuses to cover her face. She refuses to relinquish that shred of her identity and her cultural heritage. Ultimately she rejects these cultures as uncivilized. Her brushes with them show her little to admire and nothing to emulate.

But as she travels East, something interesting begins to happen to Mrs. Bruce. She begins to *notice* cultures more consciously; she expresses an interest in them totally lacking in her descriptions of her previous encounters. At first, she notices the most obvious and outward elements of change: "What struck me as the most remarkable sensation of my flight was the changing colour of man. Each time I landed, people's faces were darker. Soon they would be almost black. I thought how interesting it was going to be to see them gradually turning yellow as I journeyed northwards to China and Japan" (86). As her narrative demonstrates, however, she still lacks any sense of real involvement or engagement with these cultures.

She finds behavior interesting, also, and while still from her decidedly British bias, she tries to understand it. Her reaction to Calcutta shows both the colonial attitude of Hon. Mrs. Victor Bruce, part of the Empire, and the more alert—if naive—attention of Mary Bruce: The Hon. Mrs. tells us, "As we drove through the crowded suburbs of Calcutta, numbers of the natives leered insolently at us. There had been much uprising lately, and the inhabitants in the squalid part of the town had been particularly troublesome. I was told that it would be quite unsafe to walk after dark in any of the streets through which we were passing." Mary Bruce, trying to make sense of a phenomenon never before encountered, in a category not operable for her continues, "Driving down to the aerodrome through the same squalid streets the next morning, I saw a very interesting sight. All the streets were covered with hundreds of sleeping natives, huddled together in dirty looking heaps, some actually in gutters. They sleep out in the streets, I suppose, to keep cool" (88). That the people may have no homes in which to sleep does not ever occur to her, anymore than that their "troublesomeness" may be the logical result of desperation and justifiable anger.

Not until she reaches China do we find Mary Bruce participating in the discovery of customs and habits of thought foreign to her own. She attends a Chinese wedding even though she's tired from her travels, for she realizes, "I knew the experience alone would be worth it" (135). And if she still displays some lingering bits of Western arrogance, for example offering to trade a broken propeller for an old man's pigtail—"for I was sure I could sell it for a fortune in America" (137)—she also learns a positive lesson in China: "Although I had been in China less than a week, I had been obliged to revise my original ideas of its people, for I, like a great many others who have not visited the East, had a queer idea of the Chinese. Witness our conviction that to make a play really gruesome in England, it is only necessary for a Chinaman to walk across the stage during the first act" (148). This passage constitutes Mary Bruce's first verbalized realization that Western perspectives might be inaccurate.

By the time she reaches Japan, Mary Bruce displays much more openness to the people and ideas around her. On her first night in Tokyo, an earthquake rocks the island and causes severe damage. After touring a particularly hard-hit village the next day, she discerns, "The indomitable spirit of the Japanese, which always shows so strongly under adversity, was more than ever apparent in this little town" (182). Mary Bruce has no trouble recognizing and respecting courage when she sees it. She even gives speeches to raise money for the quake victims. And in a spontaneous gesture of cultural respect, on being the first foreigner to be awarded the Medal of the Empire Aviation Society, she follows the established custom of Japanese aviators and makes an offering at the Meiji Shrine (179).

Mary Bruce has traveled a long way from the beginning of her journey, in distance and in outlook. The Mary Bruce we come to know through the pages of this book remains an observant, thoughtful participant in her adventure. She learns from it, and if she still carries with her the vestiges of British colonialism and aristocratic privilege, we as readers come to realize just how strong the bonds of societal expectations and conventions were for her. This should be no surprise, for both of the author's personae possess strong voices; the Hon. Mrs. Victor Bruce speaks with the authority of the designated author, while Mary Bruce speaks from a compulsion that will not be denied. We may be tempted to say the two voices mutually exclude each other, but ultimately, we know, both voices emanate from the same person; both fulfill facets of the author's role. Mary Bruce *needs* Hon. Mrs. Victor Bruce in order to hold on to the respectability that enables her to live with the freedom she has without the censure that surely would otherwise come to her.

As Mary Bruce knew, society judges harshly those women who step beyond its parameters of acceptable behavior. But Mary Bruce could not be silenced; she told her story—the story of a woman in love with adventure, unafraid to grasp it where she might. Mary Bruce possessed the courage, the zest for life, and the knowledge that allowed her—as Emily Dickinson allows of poetry—to "tell the truth, [even if she had to] tell it slant."

5

Pauline Gower and Dorothy Spicer: A Shared Autobiography

The very term itself, "shared autobiography,' seems an oxymoron. After all, "autobiography" means "the story of one's self," the singular self, the individual self. Never mind that life seldom arranges itself that way, that few of us are ever totally "selves" completely devoid of the influence of others. In fact, philosophers and psychologists alike attest that our concept of self hinges to a great extent on the image of our persona we see reflected back from others. In real life, it seems, Self and Other become inextricably interwoven. The threads refuse to neatly untangle or to present a pristine story. So much for the purity of the literary genre.

The document before us, *Women With Wings* (1938), may offer evidence that the genre of autobiography sometimes proves too restricting to be accurate, for this book breaks many of the conventions we have come to expect from ordinary autobiographical writing. First of all, unlike most of the autobiographies we encounter in this volume, this document chronicles no record-breaking flights or long-distance voyages; it describes no exotic real estate or daring tests of endurance. Instead, it concentrates on the daily existence of two working pilots who successfully operated an air taxi/air circus business during the height of the Great Depression in England. What makes this chronicle noteworthy, aside from its record of commercial aviation in its infancy, is its record of two women facing tremendous odds to create a successful business in an occupation societally associated with male prowess. Gower delineates the difficulties they had to overcome: a societal age bias, for together her and Spicer's "total ages did not amount to more than forty years" (25), their corresponding lack of experience in conducting a commercial aviation business, and most strongly a societal gender bias. "It was not easy at first, nor in fact has it ever been easy—this life we have chosen. No one would take us seriously. People ragged us and did their best to hinder our training, and frequently when we were on the aerodrome did we overhear some such remark as : `What do those bloody women think they are doing here?'" (25). Perhaps more than the record setting, long distance women flyers, Gower and Spicer had to deal with that prejudice. They were, after all, living in the everyday world of aviation instead of the rarefied atmosphere of highly publicized stunt or record flights. Their rewards would not be shaped by prize money, but by the far more threatening reality of business success. They were, in very real and specific dimensions, rivals of the ordinary working male aviator. In a very explicit way, they represented more of a challenge to male domination of aviation than did the more flamboyant Amy Johnsons, Jean Battens, or Amelia Earharts of the aviation world. Thus, male reactions to Gower and Spicer tended toward the pointed.

Gower recounts in her narrative an especially vicious trick played on them by an aviator "friend" who convinced them to disassemble their engine in an effort to solve what they later discovered to be a simple and easily recognized problem. Expressing to the reader her anger over this intentional deception, Gower mentions in passing that while many aviators "had been unhelpful and unkind mostly on account of the fact that we were women, and a lot of prejudice existed at that time against women pilots" (66), this was the worst "dirty trick" they had encountered. In addition they had to face the pundits assuring society that women were not suited to highly technological careers of any kind. Nevertheless, Gower and Spicer give us evidence to the contrary, in strong female voices that shout of their cooperation, their competence, and their shared success.

More importantly—as the genre goes—their two voices together tell us the story. One cannot understand one voice without hearing the other, as well; together they provide a multi-dimensional portrait of early aviation and young aviators. Gower carries the narrative, tells the stories, creates the dominant persona, and in Spicer's words "present[s] the play"; Spicer takes us backstage to a strangely familiar, vaguely Shakespearean realm and shows us the props, as it were. It is she who takes us to the book's conception: "We would pool our memories, but who was to be the `I' and who would be just the `other fellow'?" (xv). She also answers that, after discussion—played in her account for the comic as an "acrimonious" argument that almost dissolved their business partnership—Gower who as pilot is "therefore entitled to the last word" will become the more prevalent voice, while Spicer herself will "ring up the curtain" in the book's Prologue (xv). And this she does with a deliberately theatrical caution to the audience that, "If you find these pages foolish, forgive their folly with the tolerance of the wise for youth" (xv).

Spicer's Prologue holds other pieces of information that help the reader establish a vital perspective, for with incisive strokes she paints a picture of the iconic world of aviation. She confides, for example, that before she and Gower embarked on their aviation careers, they, too, had looked at pilots as "veritable gods, beings apart. . .clean and unspoilt by the sordid touch of earth" (xvi). Acknowledging that reality soon asserted itself, she subtly demonstrates the power of the prevailing public myth—and just as acutely demolishes it. She fosters no romanticizing of the work or the people, and thus, alerts the reader to the tone Gower's voice will take.

Ordinary people those aviators may be, but, Spicer reminds us, they are still predominantly male and the language of aviation shows that bias. Thus, even though Spicer writes of their own work in aviation and Gower recounts episodes of hostility they faced from males within the aviation world, Spicer still tells us they belong to the "brotherhood of the skies." And even though Spicer, herself a world-class mechanic licensed to build and repair planes of all kinds, acknowledges that the designers and ground crew who create the planes deserve more glory than they receive, she still calls them "the real gentlemen of aviation" and seems oblivious to the irony of describing their work as that of "the Marthas of aviation" (xvii). Both intentionally and inadvertently, then, Spicer hands us as readers keys to the book before us: We are entering a male dominated landscape, led by two women who together have mastered its secrets, adopted its code, and challenged its most sacrosanct illusions. Having thus set the stage for us, she stands aside for

Gower.

Gower's persona, much less dramatic and theatrical than Spicer's, exudes comfort with self and friendship, an impression that takes shape from the vignettes and anecdotes that comprise the major part of the book. Her voice is straightforward, addressing the reader as an equal who wants knowledge of the strange world of aviation and assuring the reader that she will "try not to disappoint" (42). For Gower welcomes the reader into a world both foreign and perhaps frightening to the nonflyer. Realizing this, she adopts as one of her purposes reassuring the reader that—much more often than adventure and danger—aviation provides the opportunity for hard work and the requirement of requisite skills; in short, aviation offers a "normal" career. She focuses on the pragmatic concerns with which she and Spicer contended as they built their business in the 1930s: the chronic need for funding, the strategies for attracting customers, the seriousness of their responsibilities, the nomadic life necessitated by their goals, the tricks of the trade that helped them survive. In fact, it is through the latter that she connects most solidly with her reader, for in describing for us the relatively simple flying techniques she learned in an air circus—in essence demythologizing the sleight of hand involved in dazzling air stunts—she lets us become an insider. In revealing to us how she knows which passengers are the most nervous and how she deals with the obnoxious among them, she identifies us as a part of the in-crowd.

Throughout the book, Gower addresses her readers in an ongoing conversation, trying hard to recreate for us the flavor of those early days of aviation for two professional women who as friends and business partners set out to make their mark on the aviation enterprise. She adds to the informality of her tone by giving frequent asides to the reader, often fondly referring to a Spicer virtually reading over her shoulder, anxious to redirect or supplement the narrative. Her portrait of their relationship, while its personal elements remain private, shows a working friendship and deep admiration between two women, for their friendship is as much a subject of the book as aviation is.

Sometimes Gower uses humor to sketch their camaraderie. Describing Spicer's cooking, which she assures us is among the best she's ever eaten—"and this is really true" (133), she nevertheless laughs at her own role as official taster who gladly lets Spicer "experiment" on her. Thus, while not teaching her anything about the technicalities of cooking, Gower assures us that she has "taught [Spicer] everything she knows about the art" (133). Sometimes she adopts a more serious tone, especially in discussing the professionalism she and Spicer share. For example, in telling us that Spicer earned all four levels of ground engineer's licenses, Gower lets her admiration show: "She thus became *the only woman in the world* to have these engineering qualifications, enabling her to build aeroplanes and engines and to approve the material required for the work. I do not think that anyone quite realizes what an achievement this is" (158). Not content merely to let the notice stand on its own, she creates an enduring moment of testimonial for her friend: "I want to put it on record, here and now, that I am proud to be associated in partnership with anyone whose `business' reliability has proved as unfailing as that of her friendship—and that is saying a great deal" (158).

Even the least astute reader cannot fail to notice, however, that the glowing terms Gower uses to describe Spicer vanish when she discusses herself as aviator or as writer. Of the former she frequently downplays her very real accomplishments, on one occasion admitting that in the wake of being welcomed as a celebrity in Colombo, she felt herself to be "an impostor—that they were really expecting someone else, and would eventually find out their mistake!" (130). On another occasion, she brushed off the credit she and Spicer received as heroines when they saved another aircraft from crashing by reporting to its ground crew that it was missing a landing wheel. Their actions had been no more than would be expected of any aviator. All of this humility grew from her conviction that, if it were to take its rightful place and provide opportunities for rewarding careers, aviation needed to be demythologized, that the public must come to realize that "there is nothing very heroic in being a pilot" (114).

Of her writing, Gower—who throughout her narrative speaks to us in a conversational tone with a lively and engaging sense of humor—derides her skill as a narrator, and refers to her writing as "ramblings." In fact, as she begins a short section which she addresses specifically to women who might be considering pursuing aviation as a career, she invites those readers with no interest in such to "pass over the next page or two without laying themselves open to the charge of `skipping,' and by just so much will they be nearer the end" (214). Far from indicating her lack of skill as a writer/narrator, this exchange with the reader demonstrates Gower's very real and sophisticated understanding of her reading audience as a collective of individuals with distinct, even idiosyncratic, reasons for addressing her text. From the beginning, in fact, her relationship with her readers displays the ease of one secure in her subject and confident of her voice. She also knows when and how to shape her content, for in telling of her days as an air circus flyer, she admits that in addition to the regular logs she and Spicer kept of their experiences, they also wrote "special log-books of our memories," not intended for public consumption, but full of insights, anecdotes, and opinions "that would probably fill many volumes" (100). Clearly, she selects the material she is willing to share with her readers and saves more private memories for her personal consumption. Which is the real Gower, the humble insecure voice, or the more outspoken one? Within the text she creates, both elements co-exist in her persona, balancing each other and giving a realistic impression of an intelligent, articulate, thinking, and complex woman.

She endows her portrait of commercial aviation with that same sense of complexity. On one hand, she seems to validate the concept that aviators inhabit a different realm of reality from the rest of us. She finds the air large and lonely, and hence—somewhat unexpectedly for a professional flyer—an unnerving scope in which to work. Her descriptions reject exhilaration—the more typically expected response—for a far less definable emotion:

> To be alone in the air at night is to be very much alone indeed. . . cut off from everything and everyone . . . nothing is `familiar' any longer . . . . I think that unfamiliarity is the most difficult thing to face; one feels rather like Alice in Wonderland after she has nibbled the toadstool that made her grow smaller—and like Alice, one hopes that the process will stop while there is still something left! (42).

Even her use of ellipses to suggest her pausing, groping for words, works to complicate the sensation she describes.

Truthfully, her very words suggest, she is not always at home in the air, and indeed some of her most vivid passages describe her discomfort. For example, she recounts her own fears when during a night flight required for her commercial license examination, clouds engulfed her airplane from both above and below, thus, blocking her ability to navigate by either land or sky designations: "I was alone in the air—cut off from every living being—and from all help. I must confess that I felt small and cold and sick and anything but brave!" (44). Nevertheless, she refused to panic and in drawing that flight to a successful close, earned her commercial piloting license, as well as admiration from the licensing officials who feared her lost. Several years later, however, after an accident that left her recuperating in hospital for several weeks, she recounts her encounters with fear—"violent attacks of nerves" (189)—that plague her during her first post-accident flight. Once on the ground, the nerves dissipate, not to return, and we see the cool and competent Gower we have come to know. In recounting these experiences, Gower shows the human element of piloting, the undeniable reactions one would expect of any sensible person facing potentially fatal danger. She subtly undercuts the plaster image of bravery which denies fear; instead she describes a bravery that functions in spite of fear, if not by choice, then by necessity.

And by the way, Gower assures us the fears associated with flying are no more a woman's trait than a man's. She, from experience, recognizes the ploys by which her male passengers try to cover their fear, the nonchalant whistle which spotlights their uneasiness, the blustering about their knowledge of airplanes as they "do the most damage" to flying-wires and wings while climbing into the plane (77). How difficult and also counterproductive, she seems to say, it must be for men to be unable to show natural fear or to be convinced they must display domination over any situation, even the unfamiliar ones. The gentle ridicule in her tone conveys Gower's subtle judgment of the idiocy of the societal idea that aviation, by definition, is a man's world.

To further downplay the glamour public opinion associates with aviation and even its dangers, she allows grubby reality to intrude into her narrative. During one summer of her air circus days, a parachutist who had traveled with them dies in a fall, his parachute failing to open. She describes how they all—performers and audience joined in an undifferentiated "we"—"watched, frozen with horror" as "Evans came hurtling down without anything to check his fall—he was killed instantly." He was a friend; his death was both sudden and shocking. "But in the air-circus business there is no time for sentiment. The next day we moved on to Redcar, and although the thoughts of many of us were at Harrogate with the still, dark form we had left crumpled up on the field the night before, the show went on as usual" (122). Thus does she dispense with the presumption of glory.

Given her accounts of these far from appealing facets of aviation, we readers may legitimately ask why Gower would choose to spend her life in aviation. And she provides a well-though-through and convincing answer:

> No, it was not the novelty, and it was not the danger and the adventure (although these had their charm). It was certainly not a passing whim (if it had been the hard work would have dispelled it in a very short time!). I think there were three chief reasons for my choice of career:
>
> First, a real love for, and interest in aviation.
>
> Secondly, a determination to earn my own money and to make my career a paying proposition.
>
> Thirdly, a conviction that aviation was a profession of the future, and therefore had room to welcome its new followers (215-216).

Gower's answer belongs to a consummate professional who realistically assesses her career choice. Throughout her narrative, she acknowledges the responsibility she carries in providing as much safety, both of machine and of pilot, as possible for her passengers. In order to do so, she and Spicer constantly improved their professional credentials, Spicer obtaining all possible mechanics licenses and Gower obtaining all civilian pilots licenses, as well as those in instrument flying, radio operation, and navigation. In addition to aviation training, they also developed a keen business sense. Each business decision Gower and Spicer made was carefully weighed and pragmatically based. Their participation in the traveling air circus circuit, for example, came from their judgment that the financial solvency they would gain outweighed the danger, especially given their limited number of aviation career alternatives at the time. As soon as they could, however, they left the circuit for the relatively safer and more satisfying air taxi business, calling their company AirTrips, Ltd. (159).

With a mixture of seriousness and humor we have come to identify with her, Gower credits the success of their enterprise to a combination of their experience, their advertising and development of a clientele, and the Jubilee of King George V: "People spent their money more generously, and there was a happier and more carefree spirit abroad in the country" (161). She goes on, more seriously, to mention that the entire royal family, but especially the current King George VI and the Duke of Windsor, formerly King Edward VIII, are themselves either flyers or supporters of aviation. She even allows a touch of impishness when she suggests that the Duke of Windsor, when he was Prince of Wales, was "so truthfully 'His Royal "Highness"'" (163).

But Gower does speak pointedly to her audience about aviation and its place in the future. She notes that aviation has already become, in the span of only a relatively short time, a recognized functional and financially sound part of the world's commerce. She urges us not to overlook the dramatic acceptance of aviation, once connected in the public mind merely with war and destruction, as an "asset of peace" with the power to unite, at least physically we must add, the widely separated lands of the Empire.

She speaks seriously, as well, of what she sees as a world of opportunity for professional careers opening for women in the industries developing around aviation: "There are bound to be excellent administrative posts with large commercial companies; and probably aerodrome control towers, aeronautical instrument factories, and other auxiliaries to the industry will provide many others" (211-212). Writing, as she was, in the late 1930s, Gower cannot ignore the potential of wartime service for women aviators, especially as aerial ambulance or transport pilots. She and Spicer, together, go about demonstrating that women can successfully operate as aviation professionals in many capacities her society had not yet recognized.

If we look for a capsule version of Gower's narrative theme, one that captures both her rejection of the glory myths and her attraction to challenge, most likely we would find her following words appropriate to describe the flavor and the substance she found in her chosen career: "We are *not* born with wings! We grow them at the cost of much pedestrian hard work!" (215). Her story documents that hard work and the attendant rewards she also found. She shows herself and Spicer engaged in a career that offers them the opportunity to develop skills and a body of knowledge their own experience validates. It is a career that also puts them in charge of their own fates. In Gower's narrative, both women use those opportunities to grow strong self images, even as they forge a viable partnership in a highly technological enterprise. Gower shows us aviation as it existed as a day-to-day business at the time in which its place in societal commerce was still being assessed. Surely, hers is a timely account.

If we look more closely at this volume of reminiscences of hard work and aviation in the mid-1930s, we find a more timeless memoir of a friendship, for Gower imbues her narrative with the comfortable and mutually supportive relationship she shares with Spicer. From the asides to her readers that playfully warn of Dorothy's reading over her shoulder or Spicer's impatience to add to the story or the admiring reminders of Spicer's intelligence and skill, we sample the kind of relaxed, effortless bond that develops as two equally talented women work together. Although the words in the body of this narrative are Gower's, the essence of it contains the two of them. Gower tells the story of *their* experiences, not hers alone. But she does not have the last word.

Spicer adds an Epilogue to the volume, beginning it with the same theatrical flourish of the Prologue: "And so our story ends, although, unlike most tales, the start is merry but the end is sad" (222). From here, however, she adopts a more prosaic and far less playful tone, which is fitting for what she recounts. Telling us of plans to operate an aerial garage, she drops an enigmatic statement that "Fate robbed us both of two of our dearest possessions and took from me the full-time assistance of my partner" (222). She does not explain, and we readers are left to imagine what we will. Within the context of the book, no hint or suggestion exists of any kind that would signal changing circumstances that might result in Spicer's words. Spicer's following words expand but do not expound. She details a winter of work in a makeshift hanger which frequently flooded and a summer realization that the magic of their business was gone now that Gower was no longer a full-time partner. And so, Spicer admits, their experiment in the commercial aviation world ended, somewhat sadly and unfulfillingly, even to the reader who by now can hardly be detached from the progress of events. As readers of their book, we are left holding a mystery. Nevertheless, Spicer holds out the promise that "when the moment again seemed propitious we would begin once more" (223).

The book ends here, with Spicer's promise ringing in our ears. An irony begins to take shape at the core of that optimistic farewell. As with many forms of irony, it hinges on what as readers we can recognize as a contrast between what was imagined and what later became reality. To fully appreciate it, we must examine a piece of the volume we ignored before and then look beyond the book into the future, for together they add a dimension both informative and inadvertently poignant. First, we must look back at the Foreword, penned by Gower's and Spicer's good friend, aviator Amy Johnson, of whom we will read more in other pages of our study. She shows us a more personal side of the two women whom she considers "pioneers" of aviation because of the scope of their work and what they accomplished for aviation in general, women aviators in particular. Here, too, we find a chronicle of friendship among three women in the world of the airfield, and we find a friendship grounded firmly in mutual respect and admiration of each other's accomplishments. Johnson takes us back to 1930 when she first met Spicer, and with tongue solidly in cheek, she informs that being somewhat miffed to find another woman in what had been her exclusive domain, she hopes to find Spicer incompetent. Instead, she finds a first-rate mechanic and a lifetime friend. She tells, as well, of Gower's addition to their camaraderie as she learns flying at the same airfield. Through Johnson, we experience a weekend of gracious entertainment and fun at Gower's home and learn of other facets to their personalities. She shows us her friends, two women who are aviators to be sure, but who are also human beings with lives unrelated to flying. We come to see them in the fullness of their lives. We hear, as well, Amy Johnson who joins in their circle. Three women, great friends, and highly skilled productive aviators. They smile at us from the pages. We hear again Spicer's optimistic promise that all will be well with them.

The irony hits home, for fate intervenes. Within the decade, all three would be dead.

6

Amy Johnson's Sky Roads: A Look at Her World

This book is dedicated to all those who fell by the airwayside, for nothing is wasted, and every apparent failure is but a challenge to others.

— Dedication to Johnson's *Sky Roads of the World*, 1939

I think it is a pity to lose the romantic side of flying and simply to accept it as a common means of transport, although that end is what we have all ostensibly been striving to attain.

—Johnson, 147

As we read *Sky Roads of the World* (1939), one thing becomes clear about Amy Johnson. She knew as much about her role as writer as she did her role as aviator, and that is a considerable amount. We as readers immediately find our places in her narrative, for she tells us from the very beginning what she sees as her mission: "This book is meant to be rather a romantic story of the world's great sky routes, as seen by a pilot who has flown over most of them, than an encyclopaedic history or an information bureau—although the facts and figures quoted are as accurate as careful investigation can make them" (7). She realizes that she writes for an audience largely inexperienced in flying, even skeptical or perhaps hostile to it, at the same time becoming more dependent on the commercial advantages aviation supports and more nonchalant about its achievements. She knows her audience also hungers for sensational stories, gaining most of its information about aviation from the media, for whom Johnson harbors no misconceptions:

> Newspapers found grand material for front-page stories. The lone fight of human endurance against Nature's overwhelming odds was the favourite. Setting off unknown to face the unknown, against parental opposition, with no money, friends, or influence, ran it a close second. Clichés like "blazing trails," flying over "shark-infested seas," "battling with monsoons," and "forced landings amongst savage tribes" became familiar diet for breakfast.
>
> Unknown names became household words, whilst others, those of the failures, were forgotten utterly except by kith and kin (22-23).

In the telling of her own story with all the elements she wants to include, Johnson finds herself torn between wanting to preserve a sense of the mystique of aviation, while at the same time making her audience more knowledgeable about its realities. Her introduction carries such wording as "give life to," "make you see. . .and appreciate," the "human drama and labour" that make up the world of aviation she inhabits. Johnson is no neutral reporter talking to an impersonal audience; she is an aviation advocate and participant, an activist of sorts, writing to an audience of people she wants to reach with an intensely personal message: "I have written mostly of people I know and places I have seen, but, above all, I have tried to recreate a breath of the magic of the Air which, alas! we so often forget to notice in this present hard commercial age" (7). As writer, she brings us a gift from her heart.

Nor does she forget the reader at the end of the Introduction. She carries us with her, occasionally looking us directly in the eye and inviting us to enter her narrative. Take, for example, the passage in which she playfully urges us to "Let me fly you over this route in your own comfortable armchair. Shut your eyes and pretend you are starting with me on a flight from Cape Town to Croydon" (74). The passage that follows mixes poetic descriptions of the beauty of Africa "beckoning you to adventures undreamt-of" (75) with accurate details of airfields and geography, including available facilities and topographical information such as altitudes. On other occasions, she creates "pen-pictures" of locations over which she has flown in order to provide the reader with a sense of that airborne perspective she finds so special, as indeed have all of our flyers. She wants her reader to see and feel aviation in all its facets, guided by herself—a professional aviator who will not disguise any element of the experience.

So focused is Johnson that she even stops herself from what she sees as a digression. Her book takes the reader over the major air routes of the world, specifically those that tie the British Empire together, although she gives mention to others as well, as when she finds herself beginning to tell of her experiences at the Grand Canyon. Intended as an overnight stop on her American tour, her trip to the Grand Canyon grew into two months of flying passengers beneath the Canyon's rim "amongst its fantastic mountains and valleys" (265). Clearly profoundly impressed by its beauty and grandeur, she nevertheless recognizes the Canyon offers a different kind of flying experience than the major commercial routes she chronicles in *Sky Roads*. Thus, she reminds herself, and us, "but that is not a story for this book" (265). As exciting as those flights had been, clearly they could not be considered as part of the narrative account she is currently weaving for us. Nevertheless, she explains to the reader her decision to omit her experiences at the Canyon by embedding it in a longer discussion of the United States routes that do fit her plan, again with some of the pertinent and accurate factual details about them.

Through the pages of her book, Johnson takes us on a quick but vivid journey over the major air routes of the Pacific, Europe, Asia, the three Americas, Africa, Australia, and New Zealand. She preserves for each continent an individual flavor by providing personal observations on their approach to aviation and on the innovation they bring to the overall international aviation enterprise as it develops. Hers are the observations of a person wedded to the life of the flyer, alive with the excitement of new places and experiences, professionally aware of the challenges both met and extended for the future. More importantly, they are observations wedded to experience; those routes she discusses have been routes along which she has flown, usually in record attempts or over partially uncharted routes, in short the kind of flying she loved the most. And here we find a motif that permeates Johnson's narrative: the contrast between what she loves and what now exists.

One of the experiences she recounts offers a pertinent example of this motif. She tells us that a flight from Cairo to Cape Town in 1936 was her first flight over an organized air route, one which had regular and stable flight schedules and comfortable support facilities, and "whilst I appreciated the many blessings it provided, yet I did not particularly enjoy the experience. I could not quite shake off the feeling that I was a trespasser, and a nuisance at that" (60). She had encountered the entrepreneurs, the hotel rates, and the business-as-usual attitudes that rapidly followed the development of aviation as a relatively commonplace mode of transportation; the airplane—and the pilot—no longer exacted awe or, even, curiosity along that route. Gone are the days of the trailblazing pilot capturing new routes for others to follow. In her voice we hear a nostalgia for the mystique of flying, the mystique of adventure or danger, a mystique rapidly disappearing. We also hear a sense of personal loss, "I cannot claim to have done anything on this flight that countless others had not already done before me" (61). She loved being the vanguard. But as do all motifs, this one gains impact because it does not dominate the composition or turn into useless or bitter whining. Johnson turns to other patterns and rhythms for her narrative structure.

Assuredly, her experiences tie the air routes together, but beyond descriptions of air commerce as it existed in 1939 and beyond a recounting of her own story, Johnson provides a vivid portrait of aviation as it developed. Disclaimers from her Introduction aside, Johnson gives us a clear view of aviation history, along with the sense of adventure and beauty she cherishes. She begins at the beginning, weaving the notable people and events into the fabric of her narrative. She avoids dry statistics or disembodied facts, for Johnson seems to live the history she recounts. Indeed, she adds her own perspective to her accounts and, thus, provides for the reader further glimpses of herself even as she elucidates aviation's heritage. For example, from her we learn of the Wright Brothers and the problems of credibility their penchant for privacy at Kitty Hawk caused them: "Long and painful was the litigation that necessarily followed their efforts to prove [they were the first men in the world to fly] and to patent their designs" (12). She also points out with a gentle touch of national pride that Great Britain, not the United States, first acknowledged their claim. On a more personal level, she describes her meeting with Orville Wright in the mid-1930s by expressing a reverence for the "quiet, retiring, grey-haired man who was in truth the first conqueror of the air" (13). Under the deceptively simple guise of retelling well known facts, she has let us see her version of the human side of history; the names are people to her, the events virtually her own personal past. Her history also includes brief sketches of those who were the first to fly in France (Santos Dumont, 1907) and England (A. V. Roe, 1908, originally, but since his flight was not officially observed, the designation was given in the 1930s to Colonel Moore-Brabazon), as well as those who made other first-time flights. She expresses special pride in Sir George Cayley, a man who never saw a heavier-than-air flying machine, but "who worked out the science of flight a whole century before the rest of the world" (15). She ferries us through the War years (the only World War there had been in 1939) with the famous Aces and into the years after the War, those two decades of the glamour period for aviation, those decades of stunts and records and extraordinary flights.

We see her sense of national pride as she introduces her readers to John Alcock and Arthur Brown, British pilots who flew the Atlantic from Newfoundland to Ireland almost a full decade before Charles Lindbergh captured the glory. To be sure, she refuses to undercut the achievements of his flight, for she volunteers the information that Alcock's and Brown's flight covered a somewhat shorter distance and that there were two of them to Lindbergh's solo. Yet, nevertheless, in recounting their statistics—15 hours 57 minutes, only three miles off their destination, average speed of 122 1/2 m.p.h., 13 June 1919—she interjects, "What bare statements to describe such a stupendous feat! It was almost too great to be really appreciated, and War days had made everyone used to the unexpected" (22). And assuredly, she mourns the fact that, despite official recognition by the King of England in 1919, "to-day how few people, except historians and pilots, remember who first flew the Atlantic" (22). As her narrative continues, so do the names and the achievements she spotlights. She draws her specifics from the development of the particular route she is discussing at the moment. Thus, for example, when she talks of Australia, she brings Charles Kingsford-Smith to our attention, and of the Americas, various pilots of Pan-American Airways. Her narrative is full of names no longer familiar and achievements we see as givens, and it is full as well of Johnson's realization that the glamour years, those years that gave birth to her own brand of aviation, were drawing to a close. She clearly sees herself as witnessing the end of an era.

This impression is only heightened by the complexity of uses to which Johnson plies her own story. It is not just a straight forward, simple account. Writing in the late 1930s as an experienced professional gives Johnson an additional edge in telling her readers about aviation history, for it allows her to contrast flights from her early career, almost a decade ago, with those of the present day, to show us in specific terms how the aviation career on which she embarked a scant decade ago has changed so dramatically. She does so with an appealingly understated tone, focusing on that first world record flight to Australia in 1930:

> [A]nd except for the inconveniences of pumping all my petrol by means of an old-fashioned hand-pump, sitting in the discomfort of an open machine on a hard parachute for ten hours, instead of the four in which the "hop" could be done nowadays, being without weather reports, and knowing the whole time that if anything went wrong with my engine—more than likely—I must come down where I could, this first day's flight did not differ greatly from what it might be to-day (46-47).

With deft strokes, she delineates some of the more noticeable conditions with which the reader can identify. As Johnson realizes, these changes are clearly more likely to impress the typical reader with the advancements aviation has made in such a short period than would a discussion of engine configuration or fuselage design. When several pages later she continues by mentioning that on her original record flight to Australia in 1930, after a crash at Rangoon, the plane's wings were patched with strips torn from men's shirts and a concoction made by the local chemist [pharmacist] to substitute for the "dope and paint" usually applied to wings, she is assured of the reader's understanding of those changes.

While she understands that such descriptions have a dramatic impact on the reader, she fears they leave a false impression of unrelieved discomfort and danger. To compensate, because "I would not have you think that those early days were without their good side too," she offers us a glimpse of the prevailing view of airplanes as "novelty, something to be gazed at with awe and admiration." So true was this that she even recounts that in India, "my unexpected landing was said to have prevented a native rebellion, as it was superstitiously believed the gods had intervened" (50). The mark of those early days, she suggests, is that such a scenario really might have happened, so new was this business of flying. Yet, as readers, throughout Johnson's narrative we find that her accounts of her own experience serve an almost didactic purpose.

From our perspective, the most valuable part of Johnson's historical review of aviation, however, is its preservation of women's achievements, for more than any other of our writers, she brings her contemporaries into the spotlight with her. She steps beyond her story of herself to include what she knew about the flyers who went before her or who were her colleagues. Here, a digression of sorts seems in order, for we need to place Johnson's approach in some kind of context. Traditionally—in somewhat of an oversimplification, but a well-intentioned one— two forms of autobiography have developed in concert, each offering a slightly different emphasis in material. Some authors say to the reader, "This is my life story as I have lived it." Their major emphasis rests on the autobiographer as person; the assumption being addressed is, "People are reading this story to learn about my life," a legitimate rationale. Societal context forms the backdrop, for it is quite impossible to accurately portray one's life without some mention of the places and times in which one lived. Other authors turn the spotlight slightly aside, saying in effect, "This is the Age in which I lived, and here is how I fitted into it." Here, the assumption displays a form of tacit humility, suggesting that the surrounding period of time offers more interest than the story of one life, also a legitimate rationale. This form of autobiography pushes the background forward, making it in places the foreground, and allows the autobiographer some respite. The spotlight can be tiring. Johnson chose the second approach, rewarding her reader with small candid glimpses of familiar people under less public circumstances and, thus, giving a sense of community to the group of aviators we typically see in literary or historical isolation.

Lady Bailey benefits from this sharing, for in Johnson's voice we see her as more personable than she appeared in her own narrative. We see an added dimension. The Lady Bailey to whom Johnson introduces us takes on more grandeur, as well as more human traits; Johnson puts her achievements in an insider's perspective, one aviator discussing the accomplishments of another aviator: "Taking her time, she spent nearly a year on this flight [Croydon to Cape Town and back], and deserves her full share of credit for the fine pioneering work she did especially in flying to many parts where no aeroplane had been seen before." But she adds a more personal touch to her portrait of Bailey, "whom I know well and admire tremendously, . . a most amazing woman." Not only amazing, but also through Johnson's eyes a "most delightfully vague, will-o'-the-wisp Ladybird" who set the standards for later aviators to follow. She was the first, "and all of the rest of us have merely followed in her footsteps, although most of us have elected to stick to the straight highroad rather than wander off into the byways as she had a habit of doing" (70-71). The fondness Johnson felt for Lady Bailey shines through this slightly humorous portrait that introduces us to the gentler side of a famous aviator.

Lady Heath appears in Johnson's narrative also, but in a less personal context. One may speculate that Johnson knew of the belittling remarks about Lady Bailey which Lady Heath had included in her accounts of her solo flight from South Africa to England; indeed, it would be difficult to imagine her *not* knowing of them. Yet, in her writing she separates her personal reaction to the woman from her professional assessment of Heath's flying. In a completely different spirit than Heath displayed, she concentrates her comments on the achievement, not the woman. Thus, of Lady Heath she tells us that she "did her share" by completing that famous flight which will "go down to history as the first solo flight by a woman from South Africa to England" (71). Her terseness about the woman in no way diminished her appreciation, or her acknowledgment, of the accomplishment.

Amelia Earhart is there as well, in a passage remarkable for its poetic overtones. In the language of eulogy, Johnson writes,

> The name of yet another great pilot, Amelia Earhart, is written deep in the waters of the Pacific. Not only do they

> carry the story of some of her greatest flights, but they cherish, too, the secret of her end and enshrine forever her bright, courageous spirit (166).

With notice to the magnitude of what Earhart was attempting on her last flight, a flight that had not been successfully completed by anyone at the time, she blends recognition of Earhart's courage with what she believes all "Amelia's friends" would hold— "a secret wish she had not been quite so brave" (168). She writes as friend as well as colleague and uses a brief space within her autobiography to pause in memory: "As her friend and fervent admirer, I like to think that her spirit will inhabit those Pacific skies and protect the future Pacific airway" (169). As is fitting for an aviator she admires, the metaphorical language in which Johnson casts her last commentary on Earhart places her in the skies, where she belongs, not lost in the oceans.

Johnson documents also the place of women of in aviation. In a passage reminiscent of Virginia Woolf's famous projection in *A Room of One's Own* (1929) of what would have happened to Shakespeare's equally talented sister had she wanted to enter the world of the theater as he did, Johnson sketches the disparity between the opportunities for training and employment for men and for women. That disparity grows from two circumstances: the inexpensive training with state-of-the-art airplanes and equipment men can obtain in the Royal Air Force and the lingering prejudice women face in hiring practices and public perception. Indeed, a trace of bitterness enters her voice as she recounts to us that a prevailing prejudice keeps a woman from "being given a responsible job like that of an airline pilot. She has, however, almost as good a chance as a man to be an instructor or an air-taxi pilot—if she is good enough, whilst there are odd jobs like trailing advertisement banners, towing sail-planes, etc., for which she will probably be accepted" (26-27). The irony in the attitude that admits women can *teach* men to be pilots but not actually *be* airline pilots themselves is not lost on Johnson, nor, she wants to ensure, on her readers. For Johnson, aviation was a way of life, and to be denied entry into its opportunities a form of discrimination she found intolerable.

That is not to say that Johnson's attitude toward flying was without conflict, for in various parts of her narrative she reveals complex reactions to flying, some of which seem contradictory. In an earlier autobiographical sketch, published in a collection of narratives by famous women from all walks of life, entitled *Myself When Young* (1938) and edited by Margo Oxford, Countess of Oxford and Asquith, she admits to almost crushing disappointment at her first flight. She felt no excitement, no sensation of movement, "Just a lot of noise and wind, smell of burnt oil and escaping petrol" (141). Yet, later in that same account, she discusses her longing to fly, to gain that freedom and excitement it could bring. That same discrepancy, or maybe just dual perspective, followed her throughout her flying career. In *Sky Roads*, she can tell us of the overwhelming monotony of flying, particularly when flying over oceans where landmarks do not exist and all feeling of speed or distance rapidly disappears. On the other hand, she sees the adventure and romance of aviation, of its access to the far away and exotic, and she just as readily tells us of that facet as well.

Furthermore, she was a seasoned professional, but still remembered herself as that naive neophyte who started her long-distance career with a solo to Australia: "The prospect did not frighten me, because I was so appallingly ignorant that I never realized in the least what I had taken on, in spite of what I was told and all the awful warnings I received" (44). That flight combined both sides of Johnson, the careful planning based on available maps and information with a love of adventure fostered by a childhood steeped in mythology, fairy-tales, and adventure stories: "The dangers frightened but enchanted me" (42). She was never to escape that combination.

She knew of her own ambivalence to the lure of aviation. She said as much in the quotation with which I begin this chapter. She was forever torn between the myth and the reality, with her preference playing no favorites. In her, the romantic flyer and the professional aviator existed side by side. She is just as capable of finding flying "monotonous," as when crossing the expanses of the Atlantic Ocean, as she is of being very aware of its potential dangers:

> Hours and hours passed, with nothing to do but keep the compass on its course and the plane on a level keel. This sounds easy enough, but its very simplicity becomes a danger when your head keeps nodding with weariness and utter boredom and your eyes everlastingly try to shut out

> the confusing rows of figures in front of you, which will insist on getting jumbled together. Tired of trying to sort them out, you relax for a second, then your head drops and you sit up with a jerk. Where are you? What are you doing here? Oh yes, of course, you are somewhere in the middle of the North Atlantic, with hungry waves below you like vultures impatiently waiting for the end (108-109).

Indeed, sometimes the very hardship seems to draw her, for as she explains, she'd rather face a long flight with few stops and "practically without sleep" than a more leisurely flight, which "is more tiring by far—at any rate, for my temperament" (47). And one can't forget her almost lyrical descriptions of landscapes as they appear from the air and places rising as if from a fairy tale. (As have virtually all our aviators, she, too, describes Baghdad this way). Clearly, aviation also supplied for her a need for the exotic, the romantic, the extraordinary. But regardless of the complex motives that drew her to aviation and the multiple meanings it had for her, her attachment to aviation grew from real experience, and with that attraction grew a clear insight into the promises aviation as an enterprise had already fulfilled and would continue to fulfill.

She expresses in her book an understanding of the dynamics of aviation growth, noting the influence of commercial potential and the international political climate in shaping air routes. In addition she tells us that air routes must follow established trade routes in order to have the freight and passenger demand necessary for a growing enterprise. For example, the England to Baghdad route developed more rapidly than others because of the oil pipeline and the increasing international demand for its products. She uses Imperial Airways, the airline of the Empire, as an example of how a commercial air service comes into being and how it turns from an experimental freight carrier into a safe, efficient passenger airline. Using the England to Australia-New Zealand route as model to illustrate the process, she takes us on the stages of its development, from its first regular weekly passenger service between England and India in March 1929 (of which we learned "first hand" from Harriet Camac) to its experimental mail and freight service to Australia in 1931, and finally to a regular England to Brisbane service in 1935. At the time her book was written, a service from

England to New Zealand looked promising (54-56).

Everywhere she takes us, Johnson points out the shrinking of the world and its growing connectedness. Nowhere is this more in evidence than in her commentary on the "Airways of the British Empire," so important to her that she devotes an entire chapter to them. As was true with our other authors, Johnson found the Empire a stable part of her world view. Her visions for the future of aviation presuppose its existence, and as with the others, her Empire is decidedly Eurocentric. In her view, aviation's strength rests in its ability to bring the far-flung territories out of exile, as it were, and position them more closely to England, itself, the heart of the Empire. To that end, she discusses in some detail not only Australia and New Zealand, but also New Guinea ("The story of the opening up of New Guinea's gold fields is one of the most romantic stories in commercial aviation" 203), Canada ("The development of aviation in Canada has been so different from that in almost all other countries that it is particularly interesting to study" 209), India and Burma (each with some form of national airline), and finally, in more detail, Africa ("Africa is a country made for aviation, not only because her vast distances need rapid transport to reduce their frightening size. . .but also because Africa hides incredible beauties and wonders of nature" 227).

Turning her attention to the Atlantic, she tells us with confidence, "To-day it is an accepted fact that an air passenger service from London to New York is feasible. It is only a question of a year or two before it will be running regularly. . . .a mail service is scheduled for the summer of 1939." If from our perspective that progress seems incredibly slow and long delayed, we have only to listen to Johnson put it in a more realistic context: "This astonishing step forward in progress has been accomplished in the comparatively short space of twenty-nine years, and takes its place as one of the greatest dramas in history" (102). Twenty-nine years, she reminds us, from the Wright Brothers' first tentative flight to a powerful and growing international enterprise capable of linking whole continents in safe, passenger and freight-carrying aircraft. The future, she assures us, belongs to aviation; "if we do not recognise this we are not in step with the times" (314).

She spends considerable space at the end of her narrative talking of aviation's future, and of all the aviators we've seen, she sees most clearly the

Royal Aeronautical Society

Amy Johnson

Royal Aeronautical Society

Jean Batten

Quadrant Picture Library / *Flight International*

Dorothy Spicer

Quadrant Picture Library /*Aeroplane*

Lady Mary Bailey

AP /Wide World Photos

Mary Bruce

AP /Wide World Photos

Anne Lindbergh

AP/Wide World Photos

Amelia Earhart

AP/Wide World Photos

Louise Thaden

Quadrant Picture Library / *Flight International*

Pauline Gower

Quadrant Picture Library / *Flight International*

Lady Sophie Heath

problems and the trends that await. With a sense of assurance and confidence, she looks into the future and predicts what we will find. Her predictions are worth examining, for from her almost uncanny foresight, they resonate with what we consider familiar today. For example, she believes one of the most serious obstacles commercial aviation will face is the weather, "especially fog and icing-up" (305), which we today recognize as familiar causes of delays and, in extreme situations, crashes. Even though she sees these two elements of the weather as continuing problems, she believes that technology will help create devices that will allow airlines to function, even in inclement weather. In addition, she mentions the need to more accurately understand the dangers of "unexpected and violent down-draught resulting from some particular air conditions not understood or realised by the pilot" (310), a comment reminiscent of recent discussions about wind sheers and their role in take-off and landing crashes.

In order for aviation to operate as a profitable, serviceable enterprise, Johnson sees the necessity of three trends in the way its business is conducted. First, she finds cooperation among airlines to be more sensible than cut-throat competition. Her reasoning is simple: "there will be traffic enough for all" (307). But airlines must organize their routes. Her vision details large airplanes flying the main, more heavily demanded routes, fed by smaller "feeder" routes flown by smaller, slower, more frequent airplanes. This, in her opinion, would provide the most efficient form of service for the most customers; it has become the paradigm of modern aviation. She also sees the need for larger "aerodromes" set in open spaces with no "high obstacles" surrounding them. Her version of large seems somewhat impractical to her, for in metropolitan areas the cost of airports entailing of necessity some 700 acres with runways of 4000 feet, would be prohibitive, she believes, especially adding the cost of maintaining required hotel and ground services support. If her details somewhat miss the mark, her discussion of the trends does not. Much of what she predicted has come to pass.

As much an advocate of aviation as she is, however, Johnson makes one prediction that surprises us. Airplanes will never, she tells us, become as popular a means of transportation as the automobile. Too much operates against it. The airplane needs too much space to take off and land to make it a practical replacement for the automobile. And even if it were, the traffic problems associated with all those private planes would be "enormous": "The problems have been only partially and not very successfully solved even so far as road transport is concerned, never mind extending the troubles to the air" (311-312). Equally as importantly, however, the cost of owning and maintaining an airplane would be prohibitive for most people, including—unexpectedly for her readers—herself: "Personally, I do not now own an aeroplane because it is too expensive" (314). Even in this discussion of what will not be the future of aviation, Johnson is remarkably accurate. Her vision of the future, as it affects aviation, shines as brightly as her rendition of its past.

For, throughout this narrative—the autobiography of Johnson within her era and among the people who made that era possible— she has confronted the past with the present, the present with the future, and the future as an extension of the past. Although we have caught her longingly looking backwards at the romance and adventure largely lost to aviation as an endeavor, she ends her story with a wistful glance at the future, both what it will hold and what she hopes it will hold: "My most fervent wish is that the aeroplane will very soon have its chance to develop as an instrument to foster international trade and to enrich a lasting Peace" (314). The year was 1939.

7

Jean Batten: My Life

Jean Batten is a name which will figure for all time in history, and we can feel proud that we have lived in the same period and been the contemporaries and the witnesses of her remarkable achievements.

—Marquess of Londonderry K. G., "Foreword" to Jean Batten's *My Life*

I wanted very much to settle down in my own country and lead a calm, peaceful life, but in my heart I knew only too well that I was destined to be a wanderer.

—Jean Batten, 257

In her version of her own life, Jean Batten displays a compelling mixture of hard-nosed ambition, poetic insight, historical knowledge, and unexpected vulnerability that reveals more of the woman than just her expertise in aviation. She wrote two books, the first of which, *Solo Flight* (1934), focuses on two failed and then her successful attempt to set the women's record for a flight from England to Australia, a journey of 10,500 miles which she covered in 14 days, 22 and 1/2 hours. Published in Australia, the book remains virtually unknown in the United States, but it covers in more detail some of the same material she later used in *My Life** (1938). In this volume, she credits the success of her flight, her first real achievement as an aviator, to Lord Wakefield, whose financial patronage made the flight possible; her Gipsey I Moth plane, which performed well under all sorts of adverse situations; and her mother, about whom she writes in more detail in her second book.

For many reasons, *Solo Flight* must be described as anomalous for an aviation book, even though it entertainingly fulfills expectations for an autobiography. As one might expect for a book detailing two failed flights which occurred before the successful one, most of the narrative takes place on the ground, not in the air. Nevertheless, Batten refers to the series of attempts as her "great adventure," and she lovingly recreates the noteworthy particulars of those flights for her reader. Each of the three attempts possesses its own individuality; fittingly Batten divides her narrative into three self-contained parts, each dealing with one of the attempts.

The first attempt, begun on April 9, 1933, took Batten from England to Karachi, India, a distance of 4,800 miles—not quite half the intended journey. She describes the flight through a notice of the cities and towns she flies over, attempting through suggestions of color and shape to give the reader an impression of the beauty and also the desolation she sees around her. As she continues her story, this flight, we come to realize, was a series of near disasters capped off by a conclusive one. Plagued by intense loneliness, frequent sandstorms, and by the need for minor repairs due to small accidents, Batten chronicles for her readers the trials she faced: the cab driver in Naples who kept taking her to the wharf instead of the airport, the land crabs in the desert outside Baghdad that had just devoured a camel close to her downed plane, the rocks "native" workers had left in mounds on the runway at Jask which she discovers as she makes an emergency landing, and finally the saga of the end of her journey just outside Karachi—the sandstorm that ironically forced her to land in a field of mud, the tribesmen and women who could not understand her language, the long slow camel ride to Bela instead of to the closer Karachi where she expected to find a new propeller, and the disappearance of her plane, which had been wrapped in camel hide by the tribesmen to protect it from sandstorms, unfortunately making it invisible to the search planes looking for it. As hair-raising as these events must have been, Batten plays them for humor, a dignified humor to be sure, but nevertheless a humor born of resignation at fate's vagaries. Realizing her hope for a record was gone, Batten nevertheless hoped to finish her planned flight, only to have a connecting rod break and destroy her engine. Lucky to escape with her life, she finds her plane hopelessly damaged. As readers, we almost share her despair, especially as she tells us she was completely without financial resources. On this note, she ends the first section of *Solo Flight*.

Batten begins the second part of the book at April 31, 1934, with scant mention of the intervening months. This attempt at the record lasted only 16 1/2 hours and covered only 1,000 miles. Batten flew a plane she had recently bought, drawing for the reader a parallel between an autobiography and the plane's log book, which follows a plane from manufacture through its career and tells the complete history of its flights. The unexpected analogy provides a new perspective on the pilot's relationship with her airplane, as well as on Batten's sensibility as writer. She understands the foundation of the genre she has chosen to use.

The tone of this section differs dramatically from that of the first. Batten uses no humor at all to describe her harrowing flight that ended prematurely in Rome. She presents it as a fog-shrouded encounter as much with her psyche as it is with the elements. Ignoring advice to abandon the flight because of weather, and taking off with reduced fuel because she anticipated a short trip, Batten found herself trapped in layers of fog with zero visibility and strong head and cross winds that extended the flight time beyond what she had planned. Watching helplessly as her fuel gauge dropped to empty and visibility remained nonexistent, she describes her thoughts as nightmarish, for she expected at any moment to crash into the sea or the mountainous landscape, whichever she might find below her. Admitting that she prayed, she recounts the relief she felt when she crashed just outside of Rome, doing little damage to the plane and sustaining only a cut lip and black eye.

In her account of the flight, Batten accepts, even embraces, full blame for the crash, citing her own headstrong attitude and lack of experience. She reminds us of the need for pilots to be realistic, even in the face of ambition to reach a goal so closely within reach it almost touches one. She records for us a vow she made to herself—to learn a vital lesson from this experience.

The third section of the book, describing her successful bid for the flight record, curiously falls short of expectations. Perhaps because it documents a relatively uneventful flight—no land crabs or nightmares—it lacks the magnetism of the first two sections. Clearly written, containing descriptions of great beauty and of famous places, it provides a straight forward account of Batten's flight. But we learn more about the flyer, the writer, and the person from the first two attempts. What we *do* learn, however, from this third account is the reception Batten received from her contemporaries. She includes a selection of congratulatory telegrams at the conclusion of her narrative. Amy Johnson's is there, graciously congratulating the woman who broke her record; Mary Bruce sends her recognition, for she, too, traversed much of Batten's route; Amelia Earhart's is there, also, celebrating the achievement of another woman aviator. Other names, other dignitaries join the chorus as well, showing us as readers the impact of Batten's flight.

Solo Flight, then, provides us with a picture of a young aviator making her first impression on a world hungry for achievements in aviation. Her focus remains steadfastly on her flight, the revelations about self relevant only to her role as pilot. Four years and many flights later, Batten wrote a second book, revisiting some of this territory, but adding to it fleeting glimpses of herself as a person living a life outside the cockpit in time not always bound by records. She begins *My Life* virtually at the beginning.

Born in New Zealand "six weeks after Bleriot's historic [1909] flight across the English Channel" (15), she grew up among sports enthusiasts and the fantasy landscapes, geysers, thermal springs, and the like, of her part of the world. She tells us she learned of geography by tracing on a map her father's experiences in World War I and of adventure by reveling in the exploits of those early air racers flying between England and Australia. Indeed, one of her treasured memories as an adult was her visit with Bleriot himself and receiving from him an autographed postage stamp. She read travel books and, to the amusement of her brothers, dreamed "that some day I too would cross the sea to London" (19). She associated in New Zealand with some of the best aviators of the time: Famous airman Charles Kingsford-Smith, for example, holder of many records, took her for her first airplane ride. Not surprisingly, from her earliest years, Jean Batten loved the idea of aviation and of the adventure that inevitably accompanied it.

Despite the disapproval of her father, but aided and abetted by her mother who taught her "brave girls never cry" (18), Batten began a single-minded journey to achieve the best the world of aviation had to offer. Traveling to England with her mother ostensibly to study music in 1929,

Batten joined the London Aeroplane Club, already home at Stag Lane Aerodrome to some familiar aviators: Lady Heath, Lady Bailey, Mary Bruce, Pauline Gower, Dorothy Spicer, and Amy Johnson, among others "whom I used frequently to meet during the happy years when I flew the familiar yellow Moths at Stag Lane" (25). And so began the life in aviation Batten recounts for us. It was a life large enough to contain all the multitude of facets of Batten's personality, as she reveals them to us.

With a storyteller's skill, she takes us with her on flight after flight, breaking records and tempting fate with abandon. Frequently in telling her story, she creates and maintains suspense by taking her reader inside her mind, as her rendition of her flight over the Timor Sea demonstrates:

> About 250 miles from land the engine suddenly gave a cough. Were my ears deceiving me? I listened intently. There it was again—a sudden falter which seemed to shake the entire structure of the machine. The engine gave a final cough, then there was dead silence. A terrible feeling of helplessness swept over me as the machine commenced a slow, silent glide toward the cloud carpet. . ."Surely this can't be the end!" I thought. "No, it's impossible; there must be some way out." Almost fascinated, I watched the altimeter—5000, 4500, 4000, feet. . .This was agonizing. . . in a last desperate effort before attempting to land I opened and closed the throttle lever—without success. Suddenly, with a noise that was nearly deafening in the stillness and like a great sob, the engine burst into life again. I sank back greatly relieved, scarcely daring to breathe lest I should break the spell (96-97).

Her descriptors of the scene, for example the "final cough" and the "dead silence," work to set the mood of seriousness—even extremity—appropriate to the situation confronting us. Her steady countdown of the altimeter readings, along with her emphasis on the slowness and the silence with which the airplane moved, magnify the notion of unnatural movement. Her use of words, such as "agonizing" to describe her state of mind and "great sob" to add her personal connotation to the sound of the engine "burst[ing] into life," conveys a substantive reality to the reader. Indeed, we can almost confuse that great sob of the engine with her own outburst of relief at the sudden end to a sure disaster. We, too, sigh as Batten brings us with her and her airplane to life again.

She understands a reader's needs, sometimes moving us with foreshadows of future events, such as telling us of the conversations she used to have with fellow aviator Charles Ulm. Then just as we begin to feel acquainted with him, she admits, "Little did I realize that within six months gallant Charles Ulm was to perish in the Pacific wastes while on an attempt to span the Pacific from America to Australia" (82). Sometimes, she moves us in other directions, cutting serious discussions—for example, the following one in which she details what the job of piloting actually involves—with understated moments of humor: "My time was fully occupied steering a compass course, checking my position on the map, making up the log, pumping the petrol, and endeavouring to have an occasional sandwich or cup of coffee" (55). She uses analogy when she thinks it necessary, on one occasion explaining the red tape involved with international flying as the same kind one experiences in the more common context of ocean travel. By keeping her reader in mind, Batten weaves a coherent and rich narrative, borrowing from the pages of the literature she loves to read the techniques that allow her to place focus where it needs to be. In short, she controls the tone of her narrative with the sure touch of the storyteller.

She has a poet's touch with language, as well, which helps her create startlingly visual images for the reader as she describes landscapes over which she has flown. The passage in which she tells of a flight from Naples to Rome illustrates:

> The beauty of each successive scene, framed by the silver wings of my Moth as I looked from the cockpit, suggested a great painting, for the colours seemed too vivid and the range too great to be real. Flying along the coast I would cross occasional headlands and come suddenly upon a silvery strand of beach on which small fishing-boats would be drawn up and groups of fishermen busily engaged in spreading their nets. Little villages dotted the coast, and

> the cluster of white houses formed a striking contrast to the sapphire-blue of the Mediterranean, and the great purple, snowcapped Apennines towered away into the distance (39).

Indeed, she creates a word picture, framing it with the wings of her plane in an effort to provide a contextual perspective for the reader. She uses color to highlight the images on which our eyes would naturally rest if we, too, occupied the pilot's seat. More interestingly, the description "painting" she presents to us suggests a series of, as she calls them, "successive scenes," almost as if she sees the landscape as successive frames on motion picture film, an appropriate—if unexpected—analogy. She captures both the static nature of a painting, that moment frozen in shape and color, at the same time she suggests the dynamics of changing perspectives, the artist in motion above her subject. The result invites the reader to envision place and time and motion all at once.

Even her protests that some scenes defy description, that they virtually proclaim the impossibility of sight, hold this same visual sense of frozen motion. For example, on leaving Casablanca early one morning,

> It was impossible to distinguish anything in the blackness which enveloped the earth beneath. The sky, clear and exquisitely lovely, was encrusted with stars, like myriads of diamonds scattered at random across the vault of heaven. One by one I watched the stars fade before the oncoming dawn, and gradually the darkness gave way to a cold grey light, through which I began to distinguish the country over which I was flying (119).

In this passage, motion and light become one entity as the sky and the earth, merged in the darkness, slowly untangle.

Batten's descriptions invite a different kind of motion as well, for she displays a knowledge of and interest in archeology, history, and mythology. Thus, her descriptions sometimes take us back into time, the long stretches of time recorded in our culture: "Passing along the Gulf of Corinth, I felt that every mile over which I flew had played some important part in ancient history" (40). Aviation allowed Batten to cover more distance than can be measured only in miles. Her language evokes the stirrings of imagination about the long ago and far away, as for example, her description of flying over the "pathetic" ruins of Babylon:

> Being interested in archaeology and having read a great deal about these excavations, I longed to land and investigate at close quarters the gates with their beautiful ceramic work and the foundations of the Tower of Babel of Nebuchadnezzar's city. The winds of the desert lay wreaths on the ruins of dead Babylon, and the saying that `there is no dust-cloud in all Iraq but has in it substances that were once combined in the living person of some man or woman' must be true, for this part of the world is supposed to have been the cradle of the human race (51).

Her flights cover much of the inhabited world, and her knowledge of the ancient as well as more recent past keeps pace with them. On one occasion she draws on both her knowledge of history and her early interest in music to tell us of her visit to the "monastery at Valldemosa, where Chopin and George Sand found short-lived happiness. It was a wonderful experience to see the actual piano in the small, stone-flagged room where Chopin composed many of the preludes. I felt almost awed as I . . .realized that it was in this very room, high up in the solitude of the mountains, that Chopin composed the `Raindrop' Prelude" (205-206). In a very real sense, Batten's flights took her to other times as well as other places; they enabled her to visit history and the people who made it, people she has come to love through reading and her music.

Sometimes, the times—literal and metaphorical—in which she travels bump against one another, as she describes with a faint touch of humor: "The sun rose in a blaze of gold as I flew over the many little islands of the Aegean Sea. My thoughts of ancient Greece and the mighty Colossus of Rhodes, wonder of the Old World, were dispelled as I flew over Rhodes itself and looked down on the very modern seaplane base" (42). She has a similar experience of seeing the unexpected while flying over Rome, looking down on the excavations taking place within the city. Mussolini had ordered them screened from public view "until the work was finished and the beauty discovered revealed in full," but such screens proved no barrier for Batten, flying above them: "In one of the main streets I had seen two large coloured panels depicting the maps of the Roman Empire as it was at the height of its power and as it is to-day. The contrast, of course, was very striking, and evidently intended to create a desire in the minds of young Italians to rebuild the Roman Empire" (38). By the way, not only had she seen the unintended panels, she had also seen the motive behind them, making her one of the more politically aware of our flyers.

One cannot understand Batten's love of aviation without recognizing her love of the beauty and romance it makes possible. Alive to the sights and the ambiance around her, she revels in the changing panorama her flights provide of experience not available to those living an ordinary life: flying in "the magic light of a full moon" (42) or flying through a rainbow watching the colors "quite distinctly on the silver wings of the Moth" (41) or in her heroic terms, having a "race with the sun" as she tries to land before nightfall (35). Having described herself as a wanderer, she convinces us that flying satisfies both her imagination and her senses, and she shares the full range of her observations with us, allowing us to become vicarious wanderers with her. She recognized this side of herself, and indeed, found it almost a requirement of her profession: "In the majority of aviators I have noticed a deep artistic sense and love of beauty" (199).

But this is not to suggest that Batten was merely a witty, romantic flyer. She was also a professional aviator with knowledge and skill in maintaining her airplane and planning her flights. She respected her task as aviator, so much so that she earned, as did others of our authors, her commercial piloting license which involved both knowledge and skill in navigation, meteorology, and mechanics. "As I wished to increase my knowledge of engineering I took a course in general maintenance of aircraft and engines in the workshop of the London Aeroplane Club. . . . For several months the day used to be spent in the hangar, where attired in overalls I worked on the engines with the regular mechanics and in the evenings attended lectures and studied navigation" (27). Her flight preparations extended frequently for months with her pouring over maps, making fuel calculations, checking her engine, and the like, very much as our other authors have done. "More than once I had sat up studying charts, maps, aerodrome and meteorological data till long after midnight, determined to plan every tiny detail so that when the time came to take off I should go knowing that I had done everything possible to make this flight a success" (103). Batten knew, in this early day of aviation, that to depend solely on the knowledge of others to provide safe clearance invited disaster. She knew the necessity of possessing enough knowledge to make sound judgments of her own, and she worked to that end.

But Batten, more than the others, tells us also of her financial arrangements, for she, unlike most of the others with the exception of Amy Johnson, was more totally dependent on her own resources to finance her flights. In fact, she tells us that if she had failed her first attempt to gain her "B" license, she, for financial reasons, felt she would have had to give up flying. Particularly at the early stages of her career, she often ran into difficulty, as she explains in describing one of her first unsuccessful attempts at the England to Australia flight that was destined to bring her fame: "I was far too proud to ask anyone for help, but actually everything I possessed had gone into the flight, and I was now considerably in debt and practically penniless" (30). In this case, only the patronage of Lord Wakefield, who through his own interest in aviation had helped other flyers, allowed Batten to continue her plans.

Later in her career, Batten's flights began bringing financial rewards, always based on her own hard work. The scenario she describes after her fateful successful England to Australia flight in 1934 established a pattern that was to mark many of her flights. While the flight itself "had not been a great financial success," it opened the door for more lucrative activities, such as "lecturing, broadcasting, writing, giving passenger flights, and advertising various products" (93). Batten was, of necessity, an astute businesswoman; her developing career depended on her ability to make and use money wisely to increase her opportunities for the kind of flying she craved. In addition to her broadcast and publicity work, on occasion she discusses the need to sell one plane to help finance buying another more modern or powerful one for her flights. To a very real extent, Batten's aviation career and the records she could set were determined by the kind of airplane she could afford to fly. Many of the places she wanted to go were beyond the flight range of her planes, limiting the number of record flights she could attempt. Her business sense helped make those flights possible.

Those record attempts cannot be overemphasized, for they stand out in Batten's book as a major motivation for her flights. She was not interested in merely getting from one place to another, she wanted to get there faster than anyone else had ever done. That desire affected virtually every decision she made as flyer. For example, on her successful flight from England to Australia, her unexpected stop in Rome lasted a week, a week in which she heard the record clock keep ticking. Rather than continue her flight from that point, she flew back to England to begin the entire flight again. In explaining her reason, she adopts a position we will find more prevalent as her narrative continues: "My reason for returning to England instead of flying on was that I was reluctant to add the week spent in Rome on to my time, for I wished to make a reasonably fast flight through to Australia. It was my intention to establish at least a women's record for the journey, realizing that my aeroplane was not suitable for anything faster at this stage" (31).

But breaking women's records offered no challenge to Batten, and as soon as she could obtain more powerful airplanes, world records began to fall to her. Of her England to Brazil flight, for example, she tells us, "A wave of pleasure overwhelmed me as I realized I had lowered the record from England by a margin of almost a day, and had also crossed the Atlantic in the fastest time in history [92 days, 13 hours, 15 minutes]" (158). Record flights gave Batten a great deal of satisfaction; they formed the basis of warm memories for her, as is evidenced by her later narrative return to that flight, telling us that in retrospect it still gives pleasure: "Surely Columbus himself could not have been more pleased when in 1498 he sighted South America" (196). She records in her book the records that fall to her, pointing out that some were held by men, such as the England to Darwin, Australia, record held by H. F. Broadbent until she broke it (238) or the England to New Zealand record held by Charles Ulm, "but he had a crew" while she flew solo (251). She tells us, with obvious and candid pride, what this flight meant to her:

> This was really journey's end, and I had flown 14,000 miles to link England, the heart of the Empire, with the city of Auckland, New Zealand, in 11 days and 45 minutes, the fastest time in history. With this flight I had realized the ultimate of my ambition, and I fervently hoped that my

> flight would prove the forerunner of a speedy air service from England. . . .The triumph of the flight had been complete, and I felt a desire to stay the hand of time and enjoy to the full this hour of success (252-253).

Her successes were real, her flights *bona fide* triumphs, her joy and sense of accomplishment complete. In fact she tells us her life in aviation granted her "the greatest and most lasting of joys: the joy of achievement" (161). Her discussions of her record flights, then, always contain her description of her reaction to having accomplished what she set out to do. False modesty has no place in her narrative; instead she provides a refreshingly honest exuberance at her own courage and skill. As readers we cannot forget that Batten is a long distance flyer driven to set records, and as she carefully points out to us, she sets them solo. But as readers, we come to realize she is telling us more.

She organizes her book around her flights, and the chronology she establishes proves important for several reasons. As we would expect, it defines Batten's aviation career, placing her flights in order and showing her development as a flyer. As readers, we tend to rely on chronological organization to provide a comfortable context for either biography or autobiography; it's the way we perceive life to be lived, and, therefore, a logical way to describe it. But, chronology in this case reveals more than time line; it highlights for us the elements of Batten's persona that remain relatively the same throughout her narrative and those that change, developing as both her story and her telling of it progress. The most important elements of change in her persona are her expanding ambition and her growing candidness about admitting its pull on her. With virtually each flight, Batten's discussion of the record she wants to set or the decisions she makes that relate to a possible record become more noticeable.

One such memorable discussion deals with her ultimate decision to fly the 3700 miles from Australia to New Zealand over the Tasman Sea, against all advice, in an effort to set the England to New Zealand record, a record, by the way, that would be an addition to one she had already set on this flight. She provides for her readers the crux of the decision:

> This [opposition] was not to be wondered at, however, for

> the sudden violent storms of the Tasman were well known by all Australians and New Zealanders. No one, however, realized more deeply than I the hazards of this seldom flown sea, for I had carefully studied hydrographic charts of the South Pacific and learned of the high gale frequency and the abnormal number of cyclonic disturbances throughout each year. Several times I had crossed the Tasman by steamer, and had vivid memories of storms when I had awakened in the night and listened almost fascinated to the pounding thuds as tremendous waves shook the ship from stem to stern" (240-241).

Her reasons, however, for attempting the flight are just as clear as the problems; she wants, she tells us, "The honour of completing the first solo flight from England to New Zealand and linking those two countries in direct flight for the first time in history" (241). Knowing the dangers of the flight, knowing as well her own tiredness at the end of what had already been a long and successful flight, she decides to continue. She wins her gamble, after a long and tense flight, breaking the Broadbent and Ulm records mentioned earlier.

She also admits liking the attention she receives. The receptions and the media attention appeal to Batten, and her openness in discussing the lure of glory, as well as the attraction of the challenge, associated with record-breaking flights sets her apart from our other writers. Perhaps this love of the public commotion springs from her realization that publicizing herself makes possible her financial success; perhaps her choice of the way to earn her living springs from her love of attention. Regardless, her candor about enjoying the publicity she earns adds an appealing facet to her persona.

Through Batten's narrative, we come to realize how important those records and the way of life they mandate are to her, for she begins to tell us more about the way she shapes the personal elements of her life to accommodate them. In contrast to any of our other authors, Batten describes a rigid physical conditioning program in which she engages before each of her later flights. At first she concentrates on walking, for not only does it tone the body, it heals the mind: "As you walk all the petty little worries and doubts that sometimes crowd the mind disappear, and in

Nature's soothing presence new thoughts and inspirations come as if by magic" (217). But later, in preparation for a record attempt in a flight from Australia to England, more difficult that the reverse because of prevailing wind directions, she describes a more involved regimen: "In order to make myself specially fit for the flight I trained systematically. The training took the form of physical exercises, skipping, running, swimming, walking, and horse-riding" (263). On occasion, she credits her endurance on long, tiresome flights to the "superbly fit" condition of her body and mind (279). Although others of our authors have been athletically inclined—notably Heath, Gower, Bruce, and Johnson—Batten alone describes her workouts as systematic training for flights, rather than as a sport or a hobby. Batten's version of aviation contains a decidedly physical element, in addition to the mental one of most of our narratives. In this way, she adds a completely new element to the concept of aviation that we have seen growing during the 1930s.

But physical conditioning is not the only concession Batten makes to the demands of aviation as she wants to engage it. She tells us, in discussing her early success in aviation, that she considered whether or not marriage would be important to her. Of reports of her forthcoming marriage to a specific person, whom she does not identify by name in this narrative, she concludes, "I really felt that if I married at this stage I could not devote myself so wholeheartedly to the programme I had planned for the next few years." She knows her goals, and she knows the demands that those goals will make of her. In a clear statement of self-knowledge, she confronts the personal repercussions she has come to understand: "Now that I had tasted the fruits of success and felt the urge to rise to even greater heights, any responsibility, however light, that would in any way hinder or deter my progress was not to be considered. In short, I suppose ambition claimed me, and I considered no sacrifice too great to achieve the task I had set myself" (92-93). The entanglements of marriage would complicate her life and, more importantly, diffuse her concentration to the point that she believes them to be an impediment to what she wants most in the world.

The relationships she values tend to be with aviators, such as Charles Kingsford-Smith whose friendship lasted from the time in New Zealand when he took her on her first airplane flight until his early death in the mid-1930s over the Burmese coast. Another treasured, if brief, friendship was with the very M. Bleriot whose epic flight across the English Channel had punctuated Batten's birth in 1909. Of their meeting in 1936, when she was herself a famous aviator, Batten tells us, "M. Bleriot was most enthusiastic about my career, and we became fast friends. He liked to trace the route of my flights on a large glass globe in his beautiful *salon*" (198). As mementos of that visit, he gave her a photo of himself and "two stamps which had been specially printed to commemorate his historic flight in 1909, and which he autographed for me" (199). Clearly the life that Batten chose for herself made aviators logical friends, for she seems to have lived and breathed in the world of aviation.

But Batten maintains one close personal relationship that, rather than complicate her life, offers great solace to her. She describes her mother as "my inspiration and the guiding light of my whole life" (288). From those young adult days in New Zealand, her mother had been one of Batten's main supporters as she attempted to become an aviator. In addition to financial aid when possible, her mother provided a belief in Batten's abilities and an encouragement of her endeavors. In fact, her mother's financial aid allowed Batten to gain her piloting licenses, in the first place. As Batten's narrative records, the two became great friends, as well as mother and daughter. As evidence of the bond between them, Batten on occasion took her mother on long flying tours and on vacations, calling, for example, a two-month flight in North Africa with her mother "among the happiest [weeks] in my life" (204). One cannot help but contrast this parental relationship with the one Gertrude Bacon described about her father. While Bacon's father seemed to absorb her into his own career, Batten's mother as pictured in this narrative seems to aid and encourage Batten to succeed in a career of her own choosing, although sources other than Batten's own writings portray the relationship more negatively. Certainly, however, Batten's list of accomplishments affirm her as a woman in charge of her own vocation.

As her narrative demonstrates, she set clear goals for herself personally, but she also possessed a clear vision of what aviation means. For one thing, it means the end of isolation of a part of the world dear to her. She flies over desolate stretches of Australia and recognizes the intense isolation only the airplane can resolve, for it makes accessible those places too remote, those terrains too difficult for ordinary land travel to reach. Her record flights, whatever they have meant to her personally, have also meant more efficient transportation around the world, which she also acknowledges. For her personally, however, those flights have also meant extended periods of time spent in the loneliness only the long-distance flyer can comprehend. Many times during her narrative, she expresses a longing to see another living creature, human or not, as she flies over one kind of wilderness or another. Conversely, aviation has also brought fame and financial success.

For Batten, the airplane was the vehicle that took her to the life of which she dreamed, a life that offered her a supreme sense of achievement and worth. But more than any of these very valuable attributes, Batten's vision of aviation harks back to those elements of flying that originally attracted her. She ends her book with a description of the aviator that as surely describes her vision of herself:

> Every flyer who ventures across oceans to distant lands is a potential explorer; in his or her breast burns the same fire that urged the adventurers of old to set forth in their sailing-ships for foreign lands. Riding through the air on silver wings instead of sailing the seas with white wings, he must steer his own course, for the air is uncharted, and he must therefore explore for himself the strange eddies and currents of the ever-changing sky in its many moods (302).

This is the world Batten has joined, and it is the world that sustains her. Adventurer, yes. Professional aviator, yes. Poet, yes. Storyteller, yes. Human being, yes. All these facets circulate around each other throughout Batten's narrative. Above all, however, a strong sense of joy permeates this book, for Batten gives us the impression that she thoroughly loves the life she has created for herself.

United States Flyers

In looking at the autobiographies of the American aviators, very few patterns emerge immediately. This may be, in fact, the single most distinguishing feature of aviation as it developed in this country—its diversity. The British flyers, as we have seen, demonstrated remarkable similarity in their purposes for flying and in their specific routes. No such homogeny exists for the Americans. Perhaps the reasons behind this contrast reside in the geographical and commercial realities of the two societies.

If England had Croydon and the Empire to serve as organizing features for its aviation industry, the United States did not. Our large and varied landscapes fairly demanded widespread resources and facilities for the growth of aviation. Instead of one Stag Lane, we had many centers in which manufacturing and related enterprises could support aviation. In addition, the continental United States itself held distances that made aviation economically desirable. By making dramatic decreases in the time it took to send mail and other commodities from one coast to another, aviation proved its commercial value. Thus, we had no Empire to link together; instead we linked cities. We had no national mandate of the sort to which the British flyers responded.

For some reason, however, in the 1920s and 1930s fewer of our aviators wrote about their experiences. Only Amelia Earhart, Anne Morrow Lindbergh, and Louise Thaden of all the talented and courageous women flyers in the United States chose to publish their autobiographies during those decades. Other aviators, such as Ruth Nichols, wrote articles and other short accounts, but the urge to autobiography remained largely unattended. What the United States lacked in numbers of writers, however, it recouped in the extent of their writings. Both Earhart and Lindbergh wrote several books; indeed, Lindbergh is most often identified as a writer instead of an aviator, for she has published numerous books of essays, fiction, and journals. But as writers about aviation, these women could hardly have been more individual. Their purposes for flying, even the nature of their flights, provide a collage of the face of aviation in this country during the 1920s and '30s.

Amelia Earhart most closely resembles the British flyers, for she, too, loved long distance, record setting flights. Her territory spanned the Pacific from Hawaii to the mainland, as well as the Atlantic from the United States to Europe. In addition she became an integral part of the network of women aviators participating in national air races and working with government and industry to improve the future for women interested in the variety of careers associated with flying. Anne Lindbergh, too, participated in long distance flights that set records, as well, but her motives seem different from Earhart's. Lindbergh's flights were commissioned, not to set records per se, but to establish the parameters of a practical, efficient commercial aviation business. Her records were set almost by default, for her air route mapping flights had never before been attempted. More than any of the other aviator/authors we've read, Lindbergh pictures aviation as an escape, an opportunity for solitude with the man she loved, away from the intrusion of a news media and a public she had all too much reason to find almost sadistic in its pursuit of them. Thaden, for all her records and her obvious delight in them, portrayed a working relationship with commercial aviation, demonstrating planes and flying both passengers and cargo as the need arose. She, too, worked with government to make aviation safer and more accessible for all potential aviators, especially women.

Perhaps, conversely, this overarching diversity forms its own pattern. Perhaps the diversity natural to a country as sprawling as the United States ties these three aviators together.

8

Amelia Earhart: Searching for the Flyer

Trouble in the air is very rare. It is hitting the ground that causes it.

— *20 Hrs 40 Mins*, 225; also *The Fun of It*

Ours is the commencement of a flying age, and I am happy to have popped into existence at a period so interesting.

— *20 Hrs 40 Mins*, 310

Perhaps no one in recent history has captured the public imagination to the extent Amelia Earhart has. From the moment she flew into the media spotlight in 1928 as the first woman to fly across the Atlantic to this very moment, she has held more than her share of newspaper space. Even during her lifetime, she was a media superstar, although not always pleased with her status or comfortable with it. Nevertheless, one has only to flip through the pages of collective print media to find reference after reference to Amelia Earhart. There is a tremendous irony to be found in this.

That first flight in 1928, the one that made Earhart an instant celebrity, was a passenger flight for her. Although she was a licensed pilot, which was one of the reasons she had been invited along on this flight in the first place, and was by virtue of the flight the first woman to fly across the Atlantic, she was not allowed to touch the controls of the plane. She was to serve merely as the publicity gimmick on the flight. No other flight she ever made, until that fateful unsuccessful one over the Pacific Ocean in 1937, gained as much attention. Thus, Amelia Earhart, who through her aviation skill, business sense, and communication talent did more for aviation—and women in aviation—than virtually anyone else of her time, is remembered most frequently as the flight passenger whose mysterious death has turned imagination and hypothesis into national pastimes. The shy young woman who never felt quite at ease in the media spotlight still graces everything from careful biographies to supermarket tabloids today, almost sixty years after her disappearance. The majority of her very real accomplishments based on her courage and skill, however, are lost to collective memory.

But who was Amelia Earhart? What is her legacy to us? She left it to us in three autobiographical volumes: *20 Hrs 40 Mins* (1928) about that first flight, *The Fun of It* (1932) about her life and her aviation philosophy, and *Last Flight* (1937) a posthumous book, compiled from her log records with additional information furnished largely by publisher George Putnam to whom she was married, about her aviation career and that last disastrous flight into the Unknown. Each volume resonates with its own tones, its own purposes, but together they form a collage that defines Earhart as she wanted to be seen. Because of the obvious differences in them, these three texts must be treated separately, for they each present challenges to the reader in search of the Earhart beyond the media spotlight.

By rights, some might argue that *20 Hrs 40 Mins* should have been included in the passenger chapter of this volume, with the works by Gertrude Bacon, Harriet Camac, Marie Oge Beale, and Stella Wolfe Murray. But although she did not actually pilot the airplane across the Atlantic, Earhart undertook that flight, indeed was offered the opportunity for it, because she was a licensed pilot already attracting public notice for her flying. As a flyer herself, her perceptions of that flight differ dramatically from those of a non-pilot. They serve as a fitting introduction to the professional aviator Amelia Earhart for whom we search. The book also shows us something of the personality that made Earhart so popular with the press, for in this work she appears alternately as the shy, private young woman not expecting her fame and the assertive spokesperson using her knowledge of communication and business to sell the public on aviation; she is both the serious professional and the playful aviation enthusiast. She is also the writer. Even though the book was completed only seven weeks after the flight and shows some inconsistencies because of it, Earhart's skill with language and with vision demand attention.

As writer, Earhart begins the book by discussing it as a book, and an unworthy one at that. Rather than damning the book, however, Earhart demonstrates her connection to American literature, for she knew from her own reading one of the staples of our national writing: a typically American form of authorial humility, a convention stretching back in our literature to the Puritan conversion journals and histories and prevalent since then in our autobiographies from Benjamin Franklin's to many written today. Earhart's disclaimer of her skills as writer appears in her Foreword: "I myself am disappointed not to have been able to write a `work'—(you know, Dickens' Works, Thackeray's Works), but my dignity wouldn't stand the strain" (9). She again picks up this idea in the book's concluding pages, showing her sense of humor by applying an aviation analogy to the work she has just completed:

> Now, I have checked over, from first to last, this manuscript of mine. Frankly, I'm far from confident of its air-worthiness, and don't know how to rate its literary horse-power or estimate its cruising radius and climbing ability. Confidentially, it may never even make the take-off.
>
> If a crash comes, at least there'll be no fatalities. No one can see more comedy in the disaster than the author herself. Especially because even the writing of the book, like so much else of the flight and its aftermaths, has had its humor—some of it publishable! (280-281).

In describing the writing of her book in distinctly aviation terms, Earhart conveys to her reader the importance she places on each element, the flying and the writing, neither of which would have been sufficient for her purposes without the other. As designated historian for the flight, she kept a log of the journey and discovered as the journey progresses a growing attachment for it: "There is so much to write. I wonder whether ol' diary will hold out" (178). Quite literally for her, virtually the entire flight resided in the words she wrote down to describe it.

That Earhart considered the writing of the book(s), both the log book and *20 Hrs 40 Mins*, just as important as the flight itself can be construed from what she tells us in the remainder of her Foreword, for it provides us with a sense of the mission she adopted for herself: "I can only hope, therefore, that some of the fun of flying the Atlantic has sifted into my pages and that some of the charm and romance of old ships may be seen to cling similarly to the ships of the air" (9). She believed she had a timely message to impart to her readers in those early days of commercial aviation, and if through her wording she tended to remove herself from an active role in the success of the book, she did so to focus the attention where she wanted it to be, where she was most comfortable with its being, on aviation itself. The structure she chose for the book supports this premise even further.

20 Hrs 40 Mins is really more of a triptych than a unified volume, with each of its three sections comprising roughly a third of the book. While the sections are not labeled as such—indeed the book is presented as a collection of consecutive chapters—they do fulfill very different purposes and use distinctly different narrative techniques. The first section of the book deals briefly with Earhart's childhood and more extensively with the decade before her famous flight, when as a young woman she served as a Canadian V.A.D. in World War I and later began her aviation interests. The second section covers the flight of the Friendship, including an extensive discussion of the two-week delay the crew experienced because of weather in Trepassey, Newfoundland, at the beginning of the flight. The third section contains Earhart's views of various aspects of aviation, from training to safety and from current trends to possible future developments. Each section offers a fresh face to the reader, but even in the more ostensibly personal portions of the writing, Earhart manages to focus more directly on others than herself.

Surprisingly for a book subtitled "Our Flight in the Friendship," Earhart spends much time telling of World War I, much more so than any of our British aviators whose mention of it was at most in passing. Her motives, however, support the expenditure of time and space, for she tells why she decided to volunteer as a nurse, a V.A.D., in Canada's Spadina Military Hospital to which wounded and traumatized soldiers were sent to recuperate. She also tells us how that decision affected her life: "Four men on crutches, walking together on King Street in Toronto that winter, was a sight which changed the course of existence for me. . . . The aviation I touched, too, while approached as an entertainment was of course steeped with war. . . [The planes] were a part of war, just as much as the drives, the bandages and the soldiers" (31, 37). She tells of long hours spent listening to the stories injured aviators told and of the companionship she felt with them. They represented to her, she tells us, the romance and the adventure of aviation. Somewhat sheepishly she admits that, perhaps because of her age or the circumstances in which she found herself, the danger and the horror of wartime aviation made no impression on her. She tells of air shows as well, and of one memory that stands out from the rest of a cold winter day, "I remember well that when the snow blown back by the propellers stung my face I felt a first urge to fly" (37). Nevertheless, her personal involvement with aviation grew from her personal involvement with the war: "I have even forgotten the names of the men I knew then. But the memory of the planes remains clearly, and the sense of inevitability of flying. It always seemed to me one of the few worth-while things that emerged from the misery of war" (38-39).

The Earhart who appears in the first one hundred or so pages of her book, before the Friendship's flight, presents a picture of a young woman at once very insecure in self-concept and yet very willing, dare we say determined, to challenge societal conventions and paternal objections proscribing what she should and shouldn't be allowed to do. She mentions several occasions on which her father suddenly withdrew support, financial or otherwise, on which she had counted. One such occasion dealt with the purchase of her first airplane, when after she had already signed the purchase contracts, he reneged on a loan of $2000. Characteristically, Earhart blames herself, rather than her father, with the observation, "my salesmanship was faulty," as explanation for *his* actions. As was often the case for our women aviators, her mother provided the needed financial and psychological support. Of her mother's response to her flying, Earhart tells us, "*She* has remained sold, and it was her regret she wasn't with me on the trans-Atlantic flight, if I would go" (68).

Earhart's propensity to accept blame, or its corollary—to refuse credit, shows up in other contexts. One is a surprising lack of confidence in her own ability as a flyer. She recounts in several places that she tends to defer to men in aviation matters, "As a matter of fact, I have never asked any men to take a ride. I think I have always feared that some sense of gallantry would make them accept, even though they did not trust me. So my male passengers have always had to do the asking" (70). Even her decision to learn to fly with a woman instructor, Neta Snook, grew from this sense of inferiority, "I felt I should be less self-conscious taking lessons with her, than with the men who overwhelmed me with their capabilities" (49). Here, the issue of gender becomes inextricably linked to Earhart's feelings, and in *20 Hrs 40 Mins*, she never quite comes to terms with it.

In later passages within the book, she seems to minimize the effect of gender on women's chances at an aviation career: "Too often, I think, sex has been used as a subterfuge by the inefficient woman who likes to make herself and others believe that it is not her incapability, but her womanhood, which is holding her back" (241-242). Yet, just a few paragraphs above this statement, she admits, "Today there are ample facilities for flying instruction throughout the United States. It is, however, considerably more difficult for a woman to procure it than it is for a man. . . And it is just a little harder, too, for the woman to get this instruction at the average field than it is for the man. . .It is pretty well a man-conducted business. . . .woman is conscious that she is intruding—or something akin to that—a feeling which causes hesitation" (240, 241). She knew very well that sense of intruding, that increased difficulty that faced a woman wanting to learn to fly an airplane. The story of Earhart's own first flight has been told so often it is well known, but it deserves to be repeated in this context. On arriving for her first flight, she was told that a second pilot would ride in the first cockpit with her, for as a "nervous lady" she might try to jump out, and "There had to be somebody on hand to grab my ankle as I went over. It was no use to explain I had seen aeroplanes before and wasn't excitable. I was not to be permitted to go alone in the front cockpit" (47). Her own experiences and her own words present *prima facie* evidence of the discrepancy in opportunities for aviation careers between the two genders. Yet, she seems reluctant to admit the legitimacy of what she recounts.

Confronted by the evidence as she is, one might wonder at Earhart's hesitancy in admitting a prejudice against women in aviation, but at least two reasons for it make sense. First, the simple explanation might be that Earhart's own experience and what she saw at this time as her acceptance in aviation convinced her that gender bias played very little part in a woman's chance at a career if she truly wanted it enough to work for it. Certainly as the country's premier woman aviator in 1928, such a view makes sense. She had achieved her training and she was beginning to reap rewards for her aviation abilities, if not directly on this flight, then at least indirectly. A more complex explanation suggests that although she recognized a gender bias, her own plans for aviation and her own realization of what she could accomplish mandated that she retain a comfortable working relationship with those in power in the aviation industry. After all, she was not just a pilot; Earhart had become in the mid to late 1920s involved in the business side of aviation, working for a while as demonstrator of an experimental engine in exchange for free hangar space and using her contacts with the engine's designer and manufacturer to become a charter incorporator in Dennison Aircraft Corporation. In addition, at the time of the Friendship's flight, she held the position of vice president, and then president of the Boston chapter of the National Aeronautic Association, thus "first woman President of a body of the N.A.A." (91). Through her hard work and obvious capability, Earhart was in a position to accomplish a great deal for aviation in general and for women in aviation in particular. She would lose her voice of authority by antagonizing the establishment. Regardless of motives, however, the reader finds in this book mixed messages about perceptions of women's value in the aviation enterprise.

Any such confusion disappears, however, when Earhart discusses her own role in the Friendship's flight. As we enter the second section of the book, her narrative voice takes on a new authority and humor, creating for us a vivid memoir of an historic experience. She provides an experimental format for this section, most definably of all a section to itself, for in telling of the flight itself she uses the contents of her flight log, in most cases recounted to us "exactly as it was set down (often none too legibly!) in my log book, penciled as we in the Friendship flew northeastward, with Boston behind and Newfoundland ahead" (119). She annotates the passages from the log book, which comprise most of the description of the flight, with additional needed information, explanation, or humorous anecdote. Thus, she creates a text within a text, allowing herself as writer both the subjectivity of immediate reporting and the objectivity of the editorial account. In essence, her two personae, for they appear to be slightly different from each other, provide a fuller version of the flight than either could produce alone.

Earhart, the annotator, speaks on occasion with the eloquence of almost metaphysical imagery. For example, in describing the difficulties of maneuvering in the cramped quarters of the plane's cabin, crowded with extra tanks of gasoline, she writes, "Fortunately the physical architecture of all three members of the Friendship crew was distinctly Gothic. But even at that the two boys had to turn sidewise to get through, while I, most Gothic of all, could contrive a straight-away entrance" (125). The ethereal images of the elongated medieval statues and paintings she calls to mind provide an almost visual contrast to the metallic surfaces and confined spaces of the airplane. Throughout this section of the book, she endows the flight itself, plagued at its beginning by delays in New York and later in Trepassey, Newfoundland, with several kinds of connotative imagery, lifting it above any semblance of the ordinary into the realms of the epic. First, of the original take-off in New York, the third try, her language evokes suspense, beginning noncommittally and building to a climax: "Another day. Another start. Would it flatten out into failure like its predecessors? . . . And then, suddenly, the adventure began—the dream became actuality. *We were off!*" (117-118). She describes the take-off as a "get away" and uses theater images to enhance the drama of the alternating attempts and delays, which she calls "rehearsals" and the final take-off, which by then had become "an old act" (121).

On occasion, as annotator, she steps in to provide information the reader needs to have in order to capture the flavor of the flight. As a touch of human interest, she tells us, for example, that on the flight she carried a gift copy of Commander Richard Byrd's new aviation book *Skyward* (1928) to Mrs. Amy Guest, the woman who financed the Friendship's flight and insisted that an American woman be on board. Not only does this add faces to the organization of the flight, it also suggests the prevailing opinion among other aviators such as Byrd that her flight might, indeed, be momentous. Sometimes, however, her annotations serve more narrative purposes. She adds drama to the narrative recorded in the log book, for example, by enlarging the story of a broken spring lock on the cabin's door. To the rather sparse account in the log book, in which she recounts that during the flight she "had to hold the door shut until Slim could get back to repair it" and that as the gasoline can the door had been anchored to was being pulled out of the plane, she anchored it to herself, she adds as the annotator a passage beginning, "So, a few minutes after the take-off we nearly lost two of our crew." As she explains it, Slim came "within inches" of being sucked out of the plane and she, as she "dived for that gasoline can, edging towards the opening door," had "a narrow escape" (120). They continued the flight with the door tied by a string. Clearly the narrative gains from the magnification of danger she adds to the story. What at first seems merely worth a mention becomes for the reader the potential calamity it was.

Some of her annotations add humor to the narrative, for example, her assertion that "The gastronomic adventures of trans-oceanic flying really deserve a record of their own." In providing a glimpse at their own fare, and with her tongue placed firmly in her cheek, she demonstrates the definition of hyperbole in the "adventures" and "highlights" of their food: "Our own highlights were varied. Ham sandwiches seemed to predominate en route. At Trepassey it was canned rabbit, in London the desserts were strawberries, and home again in America chicken appeared invariably on all state occasions" (141). Later she recounts that they had failed in flight to look far enough into the basket of food prepared for them, for it contained under the substantial layer of ham sandwiches they thought their only food, a number of delicacies they unfortunately had overlooked. She uses understated humor frequently in this book, and by so doing shows a playfulness that coincides with her spirit of adventure.

Some of that playfulness is aimed squarely at her own foibles, as is shown in one engaging passage in which she turns to address the reader directly. Reproducing a facsimile of three words of text from her log book, she admits to the reader that neither she nor anyone else who's seen the page can read "that word before'stratum'," and she invites the reader to try to decipher her illegible handwriting (190). She drops all pretensions of authority in this passage and leads the laughter at her own expense. She has warned us before of the peculiarities of her log, telling us that in addition to adjusting to the motion of the plane in flight, she also had to write some of the log in darkness, so the pilot could read the plane's radium instruments at night. Using her left thumb as guide to the beginning of each new line, she used "that early training with those better-late-than-never themes" she wrote after bedtime curfew when in school. "The problem of this kind of blind stenography is knowing where to start the next line. It didn't always work" (186-187). The facsimile passage mentioned above supports her statement, as the reader can easily see.

The persona she displays in the log book is at once more terse and more uninhibited than the annotator persona. The terseness, of course, results from circumstance. The lack of inhibition comes from Earhart's sense of fun. In several passages she projects fanciful shapes onto the lakes, cloud formations, and fog patches beyond the cabin's window: "Two [lakes] are gigantic footprints; another a buffalo—another a prehistoric animal" (175). She speaks, too, of "creatures of the fog" and assures us that "Irish fogs have been described in detail, and their bilious effect, and their fairies and their little people. But on one has written of a bird's eye view of one from an imaginative eye" (180, 175-176). She sees, as well, a "true rainbow," the completed circle of delicate color just beyond the wing (180). Her later knowledge that the rainbow was caused by the propeller's action in no way lessened the thrill of seeing it. To her imaginative eye, the perspective one finds in the air transforms the ordinary into the magical.

And speaking of the difference one's perspective makes, sometimes she produces an unintentional humor, as in the following passage in which she describes some of the landscape of Canada as the Friendship flies over it:

> What makes people live on little jets of land like this one?
>
> White, white sand and curving wrinkled water, windswept and barren.
>
> I have changed my seat to a gas can, one of the two saved this morning (130).

The juxtaposition of Earhart's dismay at a landscape she finds alien and desolate contrasts with the comfortable commonplace she finds in the makeshift seating on one of the world's first trans-Atlantic flights, a situation her readers are virtually guaranteed to find equally as alien. The obliviousness on Earhart's part to the incongruity the reader sees in this situation creates humor in an otherwise serious observation. Lest we think, however, that she indicts the people themselves who live in that desolate landscape, a later passage describing the two weeks she spent in Trepassey waiting for the weather to clear settles the matter: "The cruelty of country and climate is surely a contrast to the kind hearts of the people of Newfoundland. They were untiringly good to us" (155).

Much of the log details her reaction to that delay, a growing desperation of boredom and frustration that she tries, sometimes more successfully than other times, to approach with good humor. She counts off the delay day by day, beginning with descriptions of the activities with which she and the two others tried to occupy their time. By the second week, along with the inevitable account she includes of the weather, she expresses a belief that "Job had nothing on us. We are just managing to keep from suicide" (160). Photographs she includes of the house where they stayed and the main street of the village show a windswept, flat landscape with virtually no trees or vegetation of any kind. Very little, except randomly spaced, spare and plain houses, breaks the horizon. "It is hard indeed to remain sans books, sans contact with one's interests and withal on a terrific strain" (161). So isolated were they, that Earhart didn't even know of the public attention the flight had received; she didn't think anyone would still be interested in a stalled flight already two weeks old. "The days grow worse. I think each time we have reached the low, but find we haven't. . . .We are on the ragged edge" (166-167). Through the candor of such passages, Earhart allows her reader to identify with the emotion she and the two men experienced as they waited for the opportunity to face their adventure. She invites the reader to share the drudgery of the flight, as well as the glamour.

The flight itself forms the heart of the book, and she narrates it through the pages of her flight log. Of the many impressions she captures of the flight, one stands out for its purity: "The view is too vast and lovely for words. I think I am happy—" (182). But no flight of the sort can pass without some element of suspense, and she records that as well. "[The log's] last page records that we had but one hour's supply of gas left; that the time for reaching Ireland had passed; that the course of the vessel sighted perplexed us [They seemed to have crossed a shipping lane when they should have been following it]; that our radio was useless." The word she had used in her log to describe the situation was "Mess" (191). Ultimately, finding Ireland, although not exactly the part they intended, brought the journey to its conclusion, described by Earhart with a combination of poignancy and humor all her own. The poignancy comes from the meaning she finds in the flight:

> For Bill and Slim and me it was an introduction to the Old

> World. Curiously, the first crossing of the Atlantic for all of us was in the Friendship. None that may follow can have the quality of this initial flight (198).

That famous flight that would stand in history books was more than just a statistic to the participants, and Earhart wanted her readers to know that.

The humor of the flight emerges from Earhart's description of but one of the difficulties they encountered because of landing at Burry Port, Wales, instead of Southampton as planned. No crowds awaited them. Indeed, they had difficulty attracting anyone's attention. This led to some creative solutions to mundane problems, as Earhart explains:

> [Of mooring the plane to a buoy]: We didn't doubt that tying to the buoy in such a way was against official etiquette and that shortly we should be reprimanded by some marine traffic cop. But the buoy was the only mooring available and as we'd come rather a long way, we risked offending (198-199).

Again, she uses the understated humor she prefers to describe a detail other writers might have ignored.

She completes this informal section of the account with a discussion of the celebrations and receptions awaiting the crew in London and New York. In fact, on one occasion as the crowds surged around them and they were removed to a place of safety, a warehouse, she again uses her gentle humor to describe a situation which might have been as easily frightening: "In the meantime we had tea and I knew I was in Britain" (201). She tells also of meeting Lady Heath among other women aviators at a British Air League luncheon and, tongue in cheek, of having "a real man-to-man" talk with media correspondents, which she legitimately found to be "thoroughly enjoyable" (208).

She does, however, express some discomfort with two elements of their coverage of the flight. First, she dislikes being the center of attention. Earlier in the book, she had described feeling that publicity flights made her a "clown" or a "female freak" in the public's eyes, and that because of this, she did not do many demonstration flights or air races (72, 76). She returns to the same family of metaphor to explain her reaction to the attention she was attracting: "I tried to make them realize that all the credit belonged to the boys, who did the work. But from the beginning it was evident the accident of sex—the fact that I happened to be the first woman to have made the Atlantic flight—made me the chief performer in our particular sideshow" (201). Of course, from the media's position, her presence on the flight, precisely because of her sex, was the major factor in making the flight newsworthy in the first place.

Her second objection grew from her annoyance at being labeled "Lady Lindy," as she explains. In response to the first two questions she received from a reporter on landing finally at Southampton—first, whether she knew Charles Lindbergh and second, if she thought she looked like him—she lets her reader know precisely what she thinks of the idea: "I explained that I had never had the honor of meeting Colonel Lindbergh, that I was sure I looked like no one (and, just then, nothing) in the world, and that I would grasp the first opportunity to apologize to him for innocently inflicting the idiotic comparison. (The idiotic part is all mine, of course)" (202). Nevertheless, the label would follow her for the rest of her flying career. The physical comparisons and the circumstances of their flights were too striking to be ignored by the public or the press. Whether she was comfortable or not, Amelia Earhart had become a public figure, as she begins to realize in this book.

The remainder of the book is taken up with miscellaneous information about flying, designed for the reader who has either never flown or who fears flying, but who is nevertheless very much interested in how aviation functions as an enterprise. It contains descriptions of the sensations of flight, the mechanics of flight, the safety of flight, the business of aviation, and the future of flight. Earhart even includes a section of the humor of flight, using many of the clippings and cartoons that proliferated in the few weeks between the conclusion of her flight and the completion of the manuscript for her book. As she reminds us, "With such excerpts, from the newspapers and the magazines of every day, one could go on endlessly, for aviation is woven ever closer into the warp of the world's news" (310). Certainly, the next decade would prove this statement, for the 1930s saw a proliferation of flights and books about flight, in which Earhart as a flyer and an author took a leading role.

For her second book, *The Fun of It* written four years later, Earhart chose a different, younger audience. While much of the content, indeed some of the very wording, remains the same, in order to reach the young people and children she wanted to address, Earhart changed her persona, as well as her emphasis in the elements of her story as she presents them. In addition, the Earhart who wrote *The Fun of It* had four more years of experience with the problems and the opportunities of her fame. She could not write the same account in the same way, even if she had tried, for the ensuing years had brought an insight and a maturity to her. Thus, the reader of both books gains a more detailed, more intensified impression of Earhart as aviator, writer, and woman.

Although much of the content of this book was covered in the first, in these pages she concentrates most on the elements of her story she believes will be more appealing to her younger audience. Hence, she spends more time detailing her childhood, which she describes as being "much like that of many another American girl who was growing up at the time I was, with just the kind of fun and good times we all had then" (3). Then she goes on to tell of a childhood very much different from most, beginning with a nod to the perceptions of childhood by declaring she was born "thousands of years ago. . .in Atchison, Kansas" (4). She credits her early life with preparing her for a career in aviation, although not directly. The circumstances of her life, touched upon only in outline in *20 Hrs 40 Mins* but more fully and lovingly explained here, provided her with a love of new and exotic places and people, for her father, a railroad man, took her on many trips as she was growing up. Her natural inclination for sports and games, and her awareness that her elders did not approve of them for girls, strengthened her for the challenges to convention she would later engage.

Much of the early portion of this book deals with Earhart's love of reading and stories, and the central position books held in her family's life. She even tells us that often, when she and her sister had cleaning chores to do, "instead of both pitching in and doing it together, one was selected to read aloud and the other to work" (6). Of her own reading, she adopts a self-deprecatory tone, but one designed to appeal to her readers, as she describes herself "a horrid little girl," and "Like many horrid children, I loved school, though I never qualified as teacher's pet. Perhaps the fact that I was exceedingly fond of reading made me endurable. With a large library to browse in, I spent many hours not bothering anyone, after I once learned to read" (5). Reading may have been a way to stay out of trouble, but for Earhart it was clearly a passion, as well.

Both parents read to Earhart and her sister, and both told wonderful stories to them, her mother of the days when she was growing up and her father elaborate continuing tales that featured himself as hero. In these reminiscences of her father, she creates a vastly different picture than she had in her first book of the unreliable man who reneged on promises to her. The father she portrays here would delight any child with his own sense of childlike fun and his willingness to be involved with childlike fantasy. She even provides a sample of the kind of continuing stories he told, never going beyond a chapter at a time, often placing himself as fictional character in typical cliffhanger situations. "My father's occasional death or his losing an arm or leg was apt to disconcert literal minded neighbor children who happened to be listening," but they provided delightful memories to the more imaginative ones such as Earhart herself (7). From the portrait she paints in this book, one would never guess the sadness and the pain his drinking and his irresponsibility ultimately cost her.

Nevertheless, even though she continues to stress the fun and the enjoyment she felt as a child, Earhart also suggests the less positive moments she experienced. She tries to establish an immediate rapport with readers who, for whatever reason, don't fit traditional expectations, for she identifies with them. She shares her own difficulties with the limits usually placed on girls as she was growing up. For example, she tells of her love of the strenuous activity which was deemed inappropriate for females, including some sports we consider rather conventional today—bicycling, basketball, horseback riding, and tennis to name some of them. "Unfortunately, I lived at a time when girls were still girls. Though reading was considered proper, many of my outdoor exercises were not" (8). But she does not simply repeat her own escapades; she mentions them for a reason. Her own experiences allow her to also discuss the lack of encouragement and training available in the 1930s for girls and young women interested in sports: "There has been much more attention paid to boys' athletics than to girls'. So much, in fact, that many boys have easy access to coaching in various games as well as track subjects, and most girls do not. Consequently, often little incentive is provided for girls to try to develop athletically and, also, little opportunity, when they wish to do so" (10). Later in the book, she expands more on this problem, especially as it relates to the kind of training women need in order to enter technical and mechanical professions such as aviation.

> Too often little attention is paid to individual talent. Instead, education goes on dividing people according to their sex, and putting them in little feminine or masculine pigeonholes . . .Girls are shielded and sometimes helped so much that they lose initiative and begin to believe the signs `Girls don't' and `Girls can't' which mark their paths. . . Consequently, it seems almost necessary to evolve different methods of instruction for them when they later take up the same subjects. For example, those courses which involve mechanical work may have to be explained somewhat differently to girls not because girls are *inherently* not mechanical, but because normally they have learned little about such things in the course of their education" (143-144).

She points out a familiarly damaging result of this rigid gender stereotyping: "It also makes men unwilling to recognize women's abilities" (146). The Earhart of this book has resolved any confusions her younger self might have felt about the effect of gender on one's opportunities.

She delineates the impediments young women face in their attempts to break out of the passive mold set for them. She begins with a basic, but often overlooked, element: "Feminine clothing consisting of skirts, and high heels (after one begins to grow up) certainly make more difficult natural freedom of movement" (10). She also points out that the flimsy fabrics that make up most women's clothing cannot stand the strain of much intense activity without tearing, in contrast to the sturdier fabrics used in men's garments. Earhart knows from experience the validity of her comments, for she has longstanding credentials of shocking people with her own clothing. "Of course, I admit some elders have to be shocked for everybody's good now and then. Doing so, sometimes is a little hard on the shockers, however. I know this for my sister and I had the first gymnasium suits in town. . . .No one who wasn't style conscious twenty-five years ago can realize how doubtfully daring we were" (11).

As an adult, her role as aviator also placed her in conflict with established conventions for feminine dress. In several places in this book and her first she justifies the trousers and leather jackets she found so comfortable as being necessary for the dirty and ungainly places of an airplane or hangar. More than once the press attention to her clothing had insisted that she do so. She tells an amusing anecdote that also indicates the seriousness with which she viewed this issue. During her student flyer days when she was still taking flight lessons and hiking in trousers and leather jacket to the airfield, she overheard a young girl asking if she, indeed, was an aviator. On being assured that, yes, Miss Earhart is an aviator, the child expressed disbelief because her hair was too long. "Up to that time I had been snipping inches off my hair secretly, but I had not bobbed it lest people think me eccentric. For in 1920 it was very odd indeed for a woman to fly, and I had tried to remain as normal as possible in looks, in order to offset the usual criticism of my behavior" (26). Knowing the strength of that kind of criticism, she often flew her plane wearing a skirt and close fitting hat instead of trousers and helmet in order to demonstrate that flying did not necessarily make women unstylish or improper. But she also fits this matter of appropriate dress into a larger framework, and one ultimately more damaging for women.

> Tradition hampers just as much as clothing. From the period when girls were not supposed to be able to do anything comes a natural doubt whenever they attempt new or different activities. Whether or not they are fitted to do what men do physically remains to be seen. Tennis, riding, golf and other sports seem not to be harming individuals who are fit, despite dire predictions to the contrary (10-11).

In Earhart's commentary we can hear echoes of the same kind of observation Gertrude Bacon made about her own experiences growing up at a time when the medical profession predicted bicycle seats would severely injure young women if they attempted to cycle. And of course, Earhart was writing at a time when that same medical profession was still arguing against higher education for women on the grounds that the more developed the female's intellect, the more atrophied her reproductive organs. The subject was far from academic for Earhart; it struck at the very heart of her own life and the lives of other young women who aspired to what was then considered the unconventional.

What she sees theoretically about the damage convention and tradition can do to women's aspirations, she fails to see in specifics, however. In this book, she provides an illustration of just that kind of oversight through a detail about the Friendship's flight that she did not discuss in her first volume. In working out the conditions under which she would be included on the flight, she was told that the pilot, Bill Stultz, would receive $20,000 and Slim Gordon $5,000. "Having established that, I was asked if I was prepared to receive no remuneration myself. I said `yes,' feeling that the privilege of being included in the expedition would be sufficient in itself" (60). The inequity, considering that her presence on the flight gave it its primary news appeal and, thus, its primary commercial value, appalls us today. But, the inequity extended even farther: "My own compensation which I had never really seriously considered was, in addition to the fun of the exploit itself, the opportunities in aviation, writing and the like which the Atlantic crossing opened up for me. Incidentally the fees from my newspaper story of the flight went back into the treasury of the enterprise" (60). Indeed, she virtually paid part of those salaries of the other two crew members. Earhart did not see the inequity, for her sense of fairness and her professionalism as an aviator convinced her that, since the men did all the work of flying while she rode as a mere passenger, the men deserved the greatest rewards. She could not see that her contribution to the flight ultimately had more value than theirs. In the public's mind and in the commercial exchanges, that flight existed *because* it carried the first woman to cross the Atlantic Ocean by air.

Earhart saw that value in global terms, even if she dismissed it in personal terms. On meeting Amy Guest, originator of the flight, after the Friendship's landing in Southampton, she tells her young readers: "More than ever then did I realize how essentially this was a feminine expedition, originated and financed by a woman, whose wish was to emphasize what her sex stood ready to do" (84). Earhart has determined that insight would not go unnoticed by her readers.

One of the ways she emphasizes the role of women in aviation and their ability to succeed at whatever they choose is to feature a chapter on actual careers currently held by women in aviation. As she lists the possibilities, the sheer number and variety of positions virtually overwhelms the reader, especially the reader of today who may not realize how involved women had become with aviation in the early 1930s:

> If I were to count only pilots, there would be not one woman discovered in the cockpit of any scheduled airplane . . . However, there are women who do earn their living by flying. They sell airplanes, they ferry planes about the country, they carry passengers, they instruct, they fly in the promotion department of a few companies who use airplanes for advertising and for transporting their executives.
>
> As to special positions, there are a number worthy of mention. Several women own or manage airports; several conduct schools, alone or with their husbands; several hold traffic positions of varying importance; one designs the interiors of passenger airliners. There are two women examiners in the medical staff of the Aeronautics Branch, Department of Commerce. A number of women are associated with journals of the trade (142),

and the list goes on, to include writers and artists, hostesses and others in fields closely related to aviation. In this book, too, Earhart is much more outspoken than she was before about the prejudices against women that still exist in some areas of aviation, specifically the training of women pilots.

To counter the impression that women *can't* pilot airplanes, she adds a chapter discussing famous women aviators of the day. In 1932, they were still an elite group, but she concentrates on the achievements of several, including both Anne Morrow Lindbergh and Louise Thaden. Of Anne Morrow Lindbergh, Earhart emphasizes her naturalness and her gentleness. Of her involvement in aviation, she discusses the variety of skills at which Lindbergh excelled, including piloting, navigation, radio operation, and photography. And always, inevitably, when one speaks of Anne Lindbergh as aviator, one must also mention Charles, for they flew together: "Important as aviation is in their lives, they cannot think of it in any such light. It is a profession and a present reality and quite as much a matter of fact as any other twentieth-century development" (175). Of Louise Thaden, Earhart emphasizes that she is "reckoned one of the ablest women flyers," recounting some of her achievements (178). She also mentions the friendship the two of them shared, with an anecdote about Earhart stopping by Pittsburgh on a flight from New York to Chicago to give Thaden, whose son had been born only a few weeks previously, a ride to the National Air Races. The doctor who had originally forbidden her to go had to admit "the trip had done her no harm when she returned" (178). In discussing the accomplishments of the women in this chapter, Earhart seeks to provide a credential for all women, the credential that assures the public of women's appropriateness in their given professions. She believes such effort on her part is necessary, for "contrary to legal precedent they (women) are considered guilty of incompetence until proved otherwise" (179).

Needless to say, that perception she finds on the public's part affects the purpose for her books. Her writings have a strong didactic streak. She wants to educate her readers, maybe even more than she wants to entertain them, although she does both with equal facility. Even the events she tells of her childhood reflect a desire to encourage or offer moral support to young readers who challenge tradition or convention. Her own tribulations serve to show her young readers that what is acceptable will change over time and that truth to self is vital for any kind of feeling of success. She uses her own experiences after the Friendship's flight to show the unexpected range of opportunity that accompanies any groundbreaking action, and she emphasizes what she has learned from her contacts with young and old alike as they have turned to her with questions about aviation. But she uses *The Fun of It* to also educate young readers about the scope and range of aviation. She includes in the book a lengthy discussion of aerodynamic theory in simple applied terms and a discussion of what flying is like, for passenger, student flyer, and professional aviator. Much of this is based on the material of the third section of *20 Hrs. 40 Mins.*, but again, she has shifted her emphasis, included more examples based on familiar analogies, and otherwise directed her commentary to young readers. She also includes in this book material based on some of the articles she wrote for *Cosmopolitan* magazine in the late 1920s, such as the chapter on meteorology and its vital role in the development of aviation.

She covers again the flight of the Friendship, for she knows her young readers are drawn to this book because of the fame that flight brought her. She omits any passages dealing with the lengthy and disheartening delay she experienced in Trepassey, instead focusing on her impressions of her actual flight. For her young readers, she includes vivid descriptions of the cloud formations and the world above the clouds. She even refers them to her log, explaining that when she rereads it, she finds she made more references to those clouds than to anything else. Her emphasis for this younger audience remains positive and visual. But she adds an extra treat for her young readers, as timely a treat as can be imagined for one interested in aviation.

Earhart ends the book with an afterthought, almost literally so, for—in a brief section written almost immediately after she completed the trip—she tells us of her "real" flight, *her* solo flight across the Atlantic from Newfoundland to Londonderry, Northern Ireland. She also gives us some background for the chapter. Heading the final chapter with a dateline, "London, May 25, 1932," she continues, "Indeed, the book itself was finished by the time I left New York. . . Here, at the request of the publishers, is a final chapter describing the flight itself—a postscript from overseas" (209). She provides a variety of motives for the flight, fun which she has emphasized from the beginning of the book, surely. But also something more serious, for she admits to a nagging need to attempt this solo flight since the days of the Friendship's journey. At last, with this solo flight, she really earns the title of the first woman to *fly* across the Atlantic. "It was, in a measure, a self-justification—a proving to me, and to anyone else interested, that a woman with adequate experience could do it" (210).

Of the flight itself, she briefly touches on some of its memorable moments, such the altimeter failing early in the flight, flying through a severe storm of an hour's duration, ice forming on the wings, flames coming from a "broken weld in the manifold ring" (216), and flying for extended periods through a "soup" of fog: "I depended on the instruments there to tell me the position of the plane in space, as under these conditions human faculties fail" (215). Running low on fuel and without the altimeter to help her judge her altitude, she finally lands in some "lovely pastures" in Ireland, "frightening all the cattle in the county, I think, as I came down low several times before finally landing in a long, sloping meadow. I couldn't have asked for better landing facilities, as far as that" (218). And as certainly as her flight has ended, so has her book. Fittingly, she has ended it with a scoop for her young readers.

The next years mark a period of aviation feats and business accomplishments for Earhart as she became a public spokesperson for aviation. Not until 1937 did she contemplate another book, again signaling a major record flight. In looking at *Last Flight*, our task as readers is complicated by the very uncertainty of whose words, whose structure, whose insights appear on the page. While the ideas may be Earhart's, and the overwhelming majority of the words may have originated from her, the uncertainty about any specific wording, any element of structure, any particular insight makes the entire document speculative for an examination of persona. For in order to make any judgments about persona, we must be able to accept as givens that the author's way of telling the story, including choices of words and structures, represents both the *what* and the *how* the author intended. Even in situations in which the text is heavily edited by someone else, we must presume the end result met the author's approval. With a posthumous text, we can make no such assumptions, even if, as in this case, we are assured by the compiler of the text—in this case George Putnam, Earhart's husband and publisher—that it is a faithful rendition of the author's intention.

Always with a posthumously completed text we must raise questions that have no possible answer. Would the author approve of the final editing? If given the chance, would the author have reworded or otherwise revised existing text? Would the author's final selection of material to include, the anecdotes and introspections, be the same as they were at the point at which the author ceased writing? Most writers know that a text is dynamic, almost a life form in its development. If asked at what point a text becomes completed, many writers would say, "Only when it is published, and changes can no longer be made." Indications suggest that Earhart might agree. From comparisons of her other texts, we know she revised with varying degrees of completeness specific elements of text from one book to the next, carefully shaping the overall impression of what she was saying for its new context. Sometimes she merely made word level changes, but at other times she selected new details to completely alter the reader's perception of the revisited passages. An example is the way she mentioned in excruciating detail her frustrating and depressing delay in Newfoundland for the Friendship flight in *20 Hrs. 40 Mins.* and the very brief and almost objective notice of it in *The Fun of It*. Perhaps by extension we may see the same process at work *within* a text. But this is highly speculative, and the main point remains, we will never know what she would have done with this text. All in all then, we cannot assume Earhart's version of the book would resemble at all the final published version as we now know it.

Nevertheless, the text—while telling us nothing definitive about Earhart's persona or Earhart as writer—does give us valuable *information* about her life between 1932 and the last days of her last flight. As long as we as readers understand what this book can tell us—and what it can't—we can still learn much from it. This book, more than the other two, recounts the details of planning that went into Amelia Earhart's flights. We learn about a series of flights—her famous Atlantic solo in far more detail than in *The Fun of It*, her flight from Hawaii to San Francisco, the one to Mexico, and the infamous last flight which was to have spanned the globe at its equator. Of course, the flights did not appear in the other books, not having yet occurred, but the way of describing the flights differs also in this book; it, too, is new. A great deal of emphasis, for example, is placed on ground level preparations, the equipment checks, and the various kinds of necessary reports, such as weather and availability of serviceable airports along the route, that were important accompaniments to these flights. Conversely, this text spends proportionately less time describing the actual flights themselves, with the exception of that last one which is covered by entries from Earhart's log, relayed to Putnam by cable, telephone, and letter from various points along her route.

This book, too, provides names of people who worked for the flights or who served as hosts along the way of the various flights. For example, here we learn of Jacques De Sibour's help in making arrangements for fuel and maps and in obtaining official clearances for the around-the-world flight. Violette De Sibour and Amelia Earhart had met and become friends after Earhart's solo flight across the Atlantic, and George Putnam had published Violette De Sibour's book, *The Flying Gypsies*. While *Last Flight* does not mention the specific connections with Violette, although it notes that she and Jacques are aviators who spend much time flying around Europe, it describes Jacques's contributions to Earhart's flight in his official capacity as representative of Standard Oil and as personal friend of the Putnams. And speaking of the Putnams, this book serves as our introduction to George Putnam, Earhart's husband of six years, who was not mentioned in her previous books, except briefly in *20 Hrs. 40 Mins.* as one of the people who chose her for the Friendship flight. In this last book he emerges as a helpmeet in the best sense of the word, unfailingly supportive of Earhart's aviation endeavors. We see briefly a more domestic side of Earhart, as well, for the text mentions that she was planning a July 4th party for all those associated with the flight as a housewarming for the California home she and Putnam were building.

Nevertheless, we also see Amelia Earhart as a professional aviator with more than breaking records in mind. The text deals in some detail with her relationship with Purdue University's developing aviation program. Clearly, this was to be an ongoing relationship which would allow Earhart the opportunity to work with young women interested in aviation careers. The text also outlines some of the research questions she sought to answer on her last flight, dealing with aerodynamics and human factors associated with long distance flying. These questions show a distinctly scientific focus and demonstrate the seriousness of the connection between Earhart and Purdue. Purdue even helped Earhart purchase the airplane she used in this flight.

But what strikes us most clearly as readers is the sheer complexity of aviation in 1937. This book details the enormous coordination that had become necessary among a wide variety of people and industries as aviation developed rapidly in the 1930s. In this respect, this text differs radically from the others. The flights described in this book take intense planning and replanning, as the arrangements for Earhart's last flight indicate and as this text spotlights. Originally planned as a flight going west around the world, a minor crash in Hawaii caused a two month delay while the plane was being repaired. Meanwhile, seasons progressed and weather patterns changed; optimal conditions changed rapidly into less than favorable ones. In short, that delay was enough to necessitate reversing the direction of the entire flight, heading it east, thereby making all the previous flight plans, official permissions, crew rosters, and weather information irrelevant. The text recounts the effort involved in remapping, retesting, rescheduling, etc., the arrangements for the new plan.

For the flight itself, which began on June 1, 1937, the text uses passages from Earhart's log, interspersed with passages of travel writing, detailing interesting local customs and anecdotes from the various stops along the route. The focus in these portions of the book tends toward human involvement more than descriptions of scenery or tourist sights. Human behavior rooted in cultural patterns, especially as it relates to aviation, creates particularly unusual pieces of information. For example, the text tells us that in India airports had no automobile parking facilities, for authorities tried to keep people away from them, unlike the United States which actively encouraged interest in aviation. And, of course, the log traces Earhart's route, the section from Karachi to Australia by now familiar to readers of our collection of autobiographies for she followed the main colonial route established by Imperial Airways, as have so many of our authors.

But the most important references in Earhart's log are references that in retrospect glare at the reader, for we know the outcome of the narrative and in hindsight understand the significance of what the text itself paints as minor. Problems began with the monsoons that hit the flight in Akyab, Rangoon, and Bangkok, and they continued, the text tells us in brief passing mentions, with problems in the "long-distance flying instruments" that developed in Java and problems with the plane's chronometers that developed in New Guinea affecting the ability to navigate using the stars. Nothing more is said of them in this text, for on July 2, 1937, Amelia Earhart took off from Lae, New Guinea, heading for Howland Island and infinity and the realm of mythology.

George Putnam ends *The Last Flight* with a facsimile reproduction of a letter in her own handwriting Earhart once sent him, explaining her reasons for undertaking a dangerous flight. We shall end her chapter with her own words from that letter, for they capture the spirit that underlies all of the achievements of the woman, the writer, and the aviator we have come to know:

> Please know I am quite aware of the hazards.
>
> I want to do it because I want to do it. Women must try to do things as men have tried. When they fail, their failure must be but a challenge to others (228).

9

Anne Lindbergh's Voyages of Discovery

For magic, unless it is written down, escapes one. . . Yesterday's fairy tale is today's fact. The magician is only one step ahead of his audience. I must write down my story before it is too late.

— *North to the Orient*, 13

Anne Morrow Lindbergh was acutely aware that she was contributing to a special time in history; she knew that the decade of commercial aviation's infancy was like no other before or since, poised as it was at that fleeting moment of time when the world forever changed. Aviation, she believed (and subsequent events have shown), recreated reality. As a part of it, joining in the development of aviation as we have come to know it, she felt the pull of imagination as it became reality; she felt the awe of one watching the face and the conceptual nature of the world change before her eyes. She experienced first hand the wonders and the challenges of aviation. As co-pilot, navigator, radio operator on flights she shared with husband Charles, she lived in the heady technical world of the airplane; fortunately for her readers, as poet and writer she saw and heard in the simplest situation a profound exemplar of human life. She saw beyond the realm of the machine itself and the maps that guided it to the reality of what she was witnessing. She called it magic:

> It was a magic caused by the collision of modern methods and old ones; modern history and ancient; accessibility and

> isolation. And it was a magic which could only strike spark about that time. A few years earlier, from the point of view of aircraft alone, it would have been impossible to reach these places; a few years later, and there will be no such isolation (10).

Her observations are grounded in cold, hard experience, for those words belong to the Preface of her book, *North to the Orient* (1935) which describes the flight she and Charles had made four years earlier studying the feasibility of a commercial air route over the Arctic Circle. Such a route would link North America and the Orient, drawing them much closer in distance than more obvious southern routes. Lindbergh recaptures for her readers the history of that search for the northern route, beckoning travelers from virtually the European arrival in North America to find a shorter journey to the rich spice lands of ancient Cathaia. Over the centuries the route proved elusive, she says, because of such obstacles as weather, unreliable transportation, the difficulty of finding supplies along the journey. Lindbergh knows instinctively, however, that securing the northern route would mean more than merely increasing the speed and convenience of trade; it would make possible the linking of the old with the new, the marriage of myth and reality. But in 1931 when she and Charles undertook the journey, no one knew whether flight over such large and isolated expanses of wilderness were possible.

Mapped over some of the most forbidding and unforgiving territory in the world, the route offered few safe harbors; in case if emergency, the Lindberghs knew rescue seemed highly unlikely. In many ways a grueling flight over barren faceless territory, it meant something much different to Lindbergh, however. As she met the people who lived in those isolated outcroppings, she saw another kind of magic. To her, then, the journey represented the resiliency of the human spirit and the warmth of recognition between people of very different cultures, living in very disparate circumstances. For Lindbergh turns the story of her flight into the combined stories of people she met in the isolated, inaccessible outposts and the more populated habitats of her journey. She traveled a distance herself to reach her epiphany.

Understandably enough, Anne Morrow Lindbergh knew from her

childhood she must become a writer. Her early diaries and letters, published in 1971 as *Bring Me a Unicorn*, record the yearnings of the adolescent and young adult to express herself in words, not just any words, but those which could touch her readers. In school, she gloried in her teachers' praise of her work, then sank into agonies over the next essay, often reworking and rewriting even the successful ones. From the very beginning she crafted her essays, perfected her style, but as a college student she felt a distinct weakness in her writing. She deplored the lack of tension or strength in her writing: "I really need more to think about and less to write about. Nothing I write has any backbone to it and it won't have until I absorb more" (U 21). Sheltered, she felt, and divorced from the world of experience, she needed a subject matter, a connection to the world of human beings, a challenge to think and feel the heartbeat of humanity. Aviation gave her a subject matter, for as she recorded in her early diaries: "I want to write about the amazing *variety* and intensity of relationships with people—the many kinds, the subtle distinctions, the infiniteness, the richness, the excitement of them" (168). She recognized that aviation could open the door to human lives, human living, for her.

As was fitting for someone of her classical education and her familiarity with humanities and learning, the young Anne Morrow first realized the symbolism of aviation, especially as it related to the new national hero Charles Lindbergh whom she met through her father's position as Ambassador to Mexico. Acutely aware of the emerging importance of aviation, she saw Charles as "the symbol of the most beautiful, most stupendous achievement of our age—as typical and as beautiful an expression as the cathedral was of the Middle Ages—or is it just personal magnetism: For everyone does feel immediately, I think, silenced and amazed at this man" (U 91). Of course, speaking in those first few months after his record flight across the Atlantic, she echoed the sentiments of an astounded nation, one ready for a hero who could charm the gods, themselves. The handsome, courageous, humble young mail pilot who challenged the Atlantic Ocean and won, captured the imagination of the multitudes in the United States and abroad. His fame shouted from every major newspaper and magazine in the world. But to Anne Morrow, as time and life would prove, he became far more than a symbol. Their lifelong partnership literally spanned the globe, and metaphorically carried them on not just a voyage of discovery, but also an odyssey through the times and events of the twentieth century.

In *North to the Orient*, Lindbergh describes her own introduction to the technical world, and in so doing also introduces her readers to the comic side of her persona. The flight they were attempting necessitated that they carry a radio, for no other way of tracking them or of obtaining needed weather information existed. At Charles's rather oblivious insistence, Lindbergh became the radio operator, a role she found not at all suited to her natural inclinations: "Now, Charles, you know perfectly well that I can't do that. I never passed an arithmetic examination in my life. . . .I never understood a thing about electricity. . ." (31-32). As she describes her efforts to learn Morse Code and the theoretical underpinnings of the radio, she superimposes her own thoughts with the words that surround her, providing the reader with glimpses at her dry sense of humor:

> "We might as well start with the vacuum tube," said our instructor.
>
> "We might as well," I echoed, as one replies to the dentist's phrase, "We might as well start on that back wisdom tooth."
>
> He began drawing hieroglyphic diagrams on the pad, and skipping through a rapid simple sketch of the theory. He was about to start on the second diagram.
>
> "Just a moment," I said. "Before you leave that, *where* is the vacuum tube?" (32).

She completes the tale of her labors by giving us the results of her qualifying examination, a result she deliberately underplays:

> With the help of all of the diagrams, my college textbooks, and my husband's explanations, I managed to walk into the examination room one very hot day. I walked out before my husband; but I did not go as fully into the "Theory of regeneration in the vacuum tube." He passed with higher marks (33).

Consistently she plays for humor the contrast between the Charles she portrays as a technological expert and the Anne she portrays as a willing, if incompetent, neophyte. Charles, she tells us, actually showed more confidence in her abilities than she herself did. She mentions the pride she felt when, in response to questions about why he would let his wife accompany him on such a dangerous flight, he replied that she was crew, not spouse. "And I felt even more flattered. (Have I then reached a stage where I am considered on equal footing with men!)" (61). She reports also that he deferred to her on questions about the radio, even though, she assures us, he knew much more than she.

Lindbergh's ability to laugh at her own foibles masks for the unobservant, however, the strength of her accomplishments. Her own description notwithstanding, she became a highly skilled radio operator, for in reality the job *did* suit her natural inclinations. In her role as radio operator she spent each flight with human communication foremost in her mind. In fact, so involved was she with sending and receiving the radio signals that forged their only contact with the earth, she counts as one of her most thrilling moments over the Atlantic in 1933 an unplanned and incidental radio contact with New York, some 3,000 miles away. The sheer distance overwhelmed her, although realistically she knew the operator in New York made such contacts every day. Nonetheless, his nonchalance in no way detracted from her excitement, or changed the experience for her. This simple event illustrates, however, a trademark of Anne Lindbergh's concerns—the link between humans, the things that make us more alike than different. It also demonstrates a trademark of her writing, as well: her understanding of the dual perspectives, indeed the multiple perspectives, inherent in the exploration she and Charles undertook.

Over and over in *North to the Orient*, using her work with the radio as a motif and the "dit-darr-dit" of Morse Code notation as a metaphor, Anne Lindbergh deals with the way contact with two strange aviators affected the people they met, and in turn, how the people they met affected the two aviators themselves. Multi-faceted discovery inevitably accompanies exploration for her. By nature of their survey assignments the Lindberghs visited the isolated outcroppings of human habitation, checking to see if aviation companies could realistically refuel there or offer customer comforts or insure flight safety. Many, if not most, of the people they encountered had never seen an airplane; some in the northern stretches of Canada or Alaska or along the interior rivers of China rarely saw outsiders at all. She details their various reactions, not as merely an unattached observer, but as an empathetic guest appreciative of the trouble to which they had gone to welcome her.

The chapter entitled "Baker Lake" illustrates the nature of her responses to the people on their voyages. Barely more than a collection of three or four houses and a trading post serving the Eskimos and trappers of the region, she found a group of people who chose to live in isolation—the priest who would spend the remainder of his life there, the trapper who lived in a body-sized hut several miles away because the land closer to Hudson Bay had become too crowded for him, the Mountie in charge of the outpost, the oldest Eskimo woman in the camp. She also found warm fellowship and sharing, even of the slim supplies on which the inhabitants were dependent for the full year between the arrivals of its supply ship.

Life, she found, took on remarkably similar shapes in Baker Lake, even in its strangeness. For example, the inhabitants received a years' collection of newspapers which they read one day at a time, "just as you do at home," albeit one full year later than the day of publication: "Like the mythical man on the star, I thought, with a mythical telescope, who, because light takes one hundred years to travel there from the earth, sees the Civil War a century after it happened" (68). So, too, she found the smile of recognition as their hosts talked of places from their youths, glad to know that Lindbergh had visited England or that she had seen country close to long-vacated homelands. Even so, she saw as well an alien existence; she knew she did not belong to the austere life Baker Lake demands: "I turned to the Sirius and said with silent passion, 'You *must* take us out'" (78).

But Baker Lake, as isolated as it is, touched the world Lindbergh knew more intimately. She records her surprise on arriving to hear that even in this summer of 1931—the height of the Great Depression—even the Eskimos were having a hard time economically, but then she says no more—until the last page of the chapter. She tells of receiving the next winter, as she sat safely at home having a leisurely breakfast, an invitation to a private showing of furs—furs from Revillon Freres, from the trappers, she knows, at Baker Lake. The scenes, the smells, the textures of the furs, all these memories remind Lindbergh that, "I had my private showing in the summer" (79).

In a later chapter, she remembers with genuine affection the Soviet zoologist and the trapper's wife with whom she shared stories of their own and her own children. Again, Lindbergh found the similarities in their experience more pertinent than the differences; they were all three mothers. The three women gathered over photographs and found connections they did not expect. "When I left, my boy seemed nearer to me *because* they had seen his picture and had talked of him. Perhaps the zoologist also felt closer to her boy, for she gave me a letter for him, to post in Tokyo. . . . (I don't feel out of place here, I thought. . .)" (140, emphasis added). While she does not go into anymore detail about the child in the photographs, this journey was completed in 1931; the child was Charles, Jr.; separating the experience and the telling of it, Anne Lindbergh lived a nightmare that to this day she has declined to discuss in detail, except in the sparse and carefully edited pages of her published diaries. Nevertheless, in 1935 she remembers fondly a connection made half a world away that helped her transcend a chasm even more profound than mere time or distance can ever create.

Many places the Lindberghs visited on this trans-Arctic journey seemed dismal by all accessible standards, some by virtue of climate and isolation, others by virtue of natural disaster, a few by all three. During their survey flight of the Orient, the Lindberghs came unexpectedly on a flood of mammoth proportions in China; they responded to the situation by flying relief and mapping missions in the flooded Yangtze River Valley until their own plane capsized in the swollen river. Only from the air could one grasp the extent of the flood; only by air could some of the devastated areas be recognized and located. "Only someone who has been there can imagine the amount of damage a flood can do" (210), especially in an area of subsistence farming in which she recognized, "there could be no waste; that people here lived literally from day to day; that there was no `extra' stored away" (211). The Lindberghs volunteered to help by mapping the flooded areas for the National Flood Relief Commission.

Anne Lindbergh recorded their experiences, not only in terms of what they did, but also in terms of what they felt; she saw also in this context a gigantic cosmic irony. During one of the medical missions Charles and two doctors flew to the region, starving people surrounded the pontoon plane with hundreds of sampans. They wanted food; the airplane had brought only medical supplies. In their desperation the people began swarming the airplane, threatening by the sheer force of their numbers to damage or destroy it, trapping the flyers in the flood waters. By use of a gun, which they shot repeatedly in the air, Charles and his passengers moved the villagers far enough away to allow the plane to take off; they, the rescuers, escaped, but they had to leave their mission unsuccessful:

> A moment before they had been down in that crowd of starving people, some of whom might live until spring; many would die before the waters receded. Now, headed for Nanking, safety, food, and shelter were as assured to the fliers as in their own homes. Separated from those desperate people below only by a few seconds in time, only by a few hundred feet in distance, they were yet irretrievably removed in some fourth dimension. The two worlds were separated by a gulf which, although not wide, was deep, perilous, and unbridgeable. At least it was unbridgeable to the owners of the sampans. The fliers had crossed over from one world to another as easily, as swiftly, as one crosses from the world of nightmare to the world of reality in the flash of waking (220).

Acutely aware of the divergent destinies of the people of whom she writes, Lindbergh must face a reality bought by science and technology in their most elemental forms: "The pull of a trigger, the press of a switch—without these, the three magicians flying back to Nanking would have been simply three people in a starving, dying, and devastated land" (221).

One more irony rests in her narrative of this experience, this irony born of nature instead of technology. The flooded area also held what most people recognized as the most beautiful pagoda in China. Flying over it, an island surrounded by water, Lindbergh finds rare beauty: "Centered like that, a gem in its frame, it gave one also an indescribable feeling of finality and peace, as though one had reached the end of the journey or come to the heart of some mystery. Its setting, also, intensified the impression of aloneness. Ringed by silence, the pagoda was. And the things that are alone and ringed by silence must be beautiful" (223-224). In the midst of death, ancient beauty; in the midst of disaster, peace: Anne Lindbergh's careful and understated juxtaposition makes more meaning than a more extensive analysis could provide. Stylistically, she records this event with the spare skilled strokes of the Oriental artist, from starkness to beauty.

Over and over in this book, Lindbergh comments on the way perspectives change from the vantage point of the air; they take on new forms, new relationships. That, too, she tells us is part of aviation's magic. At times, these changes can disorient one, as she recounts of her own experience on flying into North Haven, Maine, a family place, but "the familiar landmarks below me had no reality. It took my mind overnight to catch up again and I lost much of the usual joy of arrival. I have had this sensation in flying many times before—this lack of synchronization of the speeds of mind and body. Pessimistically I have wondered if rapid transportation is not robbing us of the realization of life and therefore much of its joy" (48). A disturbing question, surely, but Lindbergh is too much engaged in life to shy away from its transformations. Of herself and her generation she realizes, "We are still trying to look at the stamen of a flower with spectacles made to look at horizons. Our children will measure their distances not by steeples and pine trees but by mountains and rivers" (48-49). In this deceptively simple analogy, she uses the very literal perceptual changes in human sight that mark the deeper philosophical metamorphosis of vision aviation inevitably brings.

She finds such potential ultimately and unquestionably positive, for, again using an implied analogy, she recognizes that, as her plane lifts off taking her on her journey, "The different parts of the island, also, which had once been many complicated worlds, were joined together and simplified by this enveloping glance from the air" (53). In several passages within the book, she points out that the airways have no fences, no maps, no markers. The landscape beneath the plane shows no governmental boundaries. She sees one earth, alive in its diversities, but joined by its very nature. She reminds us as she ends her account of this voyage over the top of the world, "And if flying, like a glass-bottomed bucket, can give you that vision, that seeing eye, which peers down to the still world below the choppy waves—it will always remain magic" (244).

By the time she wrote her next book, Anne Lindbergh's growing insight and subtlety as a writer allowed her to create an adventure story within an adventure story. In *Listen! The Wind* (1938) she details the southern portion of the survey flight she and Charles flew around the North Atlantic and from Africa to South America, again studying the feasibility of a commercial air route. She provides a remarkable exploration of the multiple perspectives she always recognized and in so doing, changes this book from an account of an aviation adventure to an exploration of the human condition. Ostensibly the account of their flight from the African coast to South America, the book focuses on Lindbergh's journey as aviator and person through the myriad emotions surrounding that flight.

With the radio as voice and Morse Code as its words serving successfully as a unifying motif in *North to the Orient*, Lindbergh uses a similar device in *Listen! The Wind*, as its title suggests. As both voice and language, in fact virtually a character in the book, the wind directs the course of the flight Lindbergh records, metaphorically giving the necessary permissions for the flight to continue, a special flight toward home, the long survey task completed. Lindbergh made the choice in this book not to describe the entire survey itself for her readers; she had done so earlier in an article for *National Geographic* and felt no more elaboration was necessary. Instead, she focused on particular elements of the flight, elements that because of their nature made lasting impressions on her. Thus, she divides the book into two sections: "Santiago" and "Bathhurst," the first an island just off the coast of Africa, the second a town poised on Africa's westernmost point, both potential launching points for the long journey home. Lindbergh has chosen her narrative points well, for geography serves as metaphor for that emotional territory between the end of a planned adventure and the arrival safe at home.

All good narratives, even the true ones, need some form of conflict, and in adventure stories, that conflict frequently takes the form of the elements. The same holds true in Lindbergh's narrative, for the wind—so necessary for an airplane's flight—can also be the enemy. Too much wind can tear a plane apart; too little can make its take-off impossible. Lindbergh finds in her journey home the prerequisites for the perfect narrative: Dependent on the wind to lift their airplane, but held prisoner in the Cape Verde Islands and later the coast of Africa by a constant, too strong, uncooperative version of the wind—or its equally uncooperative total absence—facing the prospect of six months or more before take off will be possible, she recreates with probing intensity the growing sense of frustration, hope, disappointment, despair, relief, and joy that lead to their escape—backtracking to a destination farther away from their goal, to be sure, but one more likely to allow them to continue forward. She focuses on those frozen moments of time in which she and Charles found themselves at the mercy of the impersonal, invincible wind:

> Listening to that wind roaring above us distantly, I had a sudden feeling of panic; a sense that it was life up there hurrying by, a great stream, tumbling, turning, sparkling, a rich swift life like the packed months just behind us. . . .we were in that stream once, but now we had been tossed out of it. We here, on this island, were caught in an eddy, a backwater, out of the stream. . . (49).

The incongruity between the relative speed and distance of Lindbergh's air travel and her enforced delay in a place that shows no evidence of time's passage adds to the drama of the conflict. In keeping with this contrast, Lindbergh draws out action in her account, lengthening it to the point of virtual standstill, so that waiting, itself, takes on the characteristics of action in the narrative. Far from being passive, waiting becomes an active component of what occurs on Santiago. In much the same way we saw Earhart try with little success to come to acceptance with her delay in Trepassey, we see Lindbergh struggling with the stopover in Santiago and Bathurst. In both cases, the aviators faced their own helplessness in the face of powerful Nature. Lindbergh expresses this feeling in mythic terms as she describes her plane's landing at Praia, Santiago, amid fierce wind that swept the harbor: "The plane bucked like a horse as we passed over the valley, circled the town, and turned out to sea again. . . Perhaps my picture was wrong. Perhaps we were not riding the wind like omnipotent gods. Perhaps, instead, our plane was a tiny sliver of bark, tossed this way and that on the choppy surface of a great unruly sea of air" (10).

As the days on the island extended, the wind became a veritable howl, its sound penetrating the very bones, its presence shaping everything on the island. The islanders seemed not to notice, but to Lindbergh, "It was something of more substance than thin air, more permanent and universal than a single scarf of breeze, rippling over a hill. It was more like a great river, a wave, which drowned the hill, and sheathed the island" (40). Always, always the wind. Such strong wind would make taking off with a full load of fuel impossible. But Lindbergh had not yet heard the worst, as the following exchange illustrates:

> "When does [the wind] stop?"
>
> "Well"—[the port agent] paused to consider—"it will blow like this for six months—. . . . No—never calm—not now—not for six months."
>
> For six months that sound of a rising wave; for six months that long sigh! (73).

But Nature did not present the only challenge for Lindbergh; she also had to deal with new and alien perceptions of time, distance, importance. At the same time, she contrasts her own emotions—that internal turmoil belonging to one whose plans have been delayed—with the patient despair and courtesy of their inadvertent hosts, such as the "Chef" in charge and his young quiet wife at Porto Praia, for whom time and alternatives seem nonexistent: "She was so quiet that one forgot her presence. There was no sense of hurry, no stir of impatience in her attitude. She had simply drawn herself up into that knot of waiting, as taken-for-granted, as familiar to her as eating or sleeping" (47). That sense of resignation permeates the residents of Santiago, for they have learned, of necessity, to co-exist with the wind; they even know its various forms and habits. Lindbergh adopts that strategy in her narrative, providing in anecdotal form a treatise on the various kinds of waiting and the similarities in their quality even if their purposes differ. She even slows the rhythm of her prose, giving it an almost somnambulant quality, to emphasize the mood of this part of the trip.

So out of synchronization with the Lindberghs, the temperament of Santiago offered special challenges; it represented to Lindbergh a gulf over which she could not go: "We were separated by something else, of our own choosing, something I felt only dimly conscious of, yet I knew was there—some test of endurance, some ordeal by fire" (76). Even the airplane's departure sparked more soul searching for her, for she realized that now time has begun moving in her world again. She could not help reflecting that her time and theirs forever diverged; effectively the people of Santiago remained frozen in her mind, belonging to a particular moment, forever unchanged. "They were now part of the past and—looking down on them was looking down at life from the altitude of death. . . . They would cease to be important to me, I realized. . . .I did not want to—but I would forget them. . . .Our plane, our eyes, and, now, our minds also would turn in another direction. . ." (118). Typically, however, she hesitated to break that human contact that existed between them, and as her first act after the plane's take off, she established radio contact with her former hosts.

The second section, "Bathhurst," offers an ironic counterpoint to the first. Further away in actual distance than Santiago, the town represented a backtracking for the Lindberghs, for ultimately their escape from the winds of Santiago required them to travel significantly further east to reach their western goal. Bathhurst, a British possession in Africa, bustled with activity, and as Lindbergh noted, "Life was going on here; it meant something. Time counted; we were in the stream again" (125). But the change in their plans meant that they had added two hundred miles to their South Atlantic crossing, and an extra two hundred miles meant that gasoline supplies, weather charts, flight schedules, and the like had to be revised. Lindbergh takes her reader through suggestions and snippets of conversations, weighing the options, charting the possibilities. And finally, finding out the other problem—no wind, no wind at all except maybe "a good breeze in the morning—at daybreak, that is" (139). Lindbergh describes her dawning realization: "Take-off and landing, wind and weather—the problems were still there" (140). And they settle down to wait for some wind.

Lindbergh details the decisions they made about what to take and what to leave, how to lighten the plane even more, whether to jettison some of the emergency supplies in order to facilitate take-off in the shallow wind. She tells also of their attempted take-offs, letting us hear the sounds of the waves slapping against the pontoons and the roar of the engine as it strained to lift the plane. She shares with us something else as well, something perhaps of more value to her; as she waited for the wind to rise, she read poetry: "It might help, too. Poetry did, sometimes, filling up the mind. I might need it tonight, bobbing up and down under the stars, or even plunging ahead through the dark sky, over the dark ocean—if we got off—" (204). For the remainder of this section of the book, a line from Humbert

Wolfe's poem, "Autumn Resignation," serves as motif. In fact, it is the line that gives the book its title. The poetry works, and as the plane ascends, Lindbergh exalts, "But now, we are free. We are up; we are off" (217), heading for South America and ultimately home.

Lindbergh's description of the ocean crossing, itself, takes only a scant forty pages of her text, and it covers her perceptions as she flies in the cockpit behind Charles's with radio equipment and earphones. Her earphones link her to the world, but to a strange and chaotic world: "Here all was still, snug and ordered. My husband sat calmly at the controls; I was curled up in my seat. . . . But out there, in that outer cosmos, so it seemed to me listening, worlds were crashing; planets were breaking up" (231). Into that static-filled night, the familiar "dit-darr-dit" of Morse Code reassures her: "A path had opened up through the night" (234). She had established again a link with human beings. Ten hours out at sea, she contacted Santiago, she tells us, for the people there had touched her life. Ironically, another of those radio links proved to be a reporter trying to obtain the world's first ocean to land interview in Morse Code; Lindbergh tapped back that she was too busy to respond, but even the intrusion of the reporter could not quell her enthusiasm for the messages she could send and receive. So much communication, she seems to say to us, from so many far off points of the compass. And finally, sixteen hours after they left Bathhurst, the Lindberghs landed in Natal, their South American destination. Ironically, they land with a good, strong wind, the kind that makes take-offs easy, the kind that had eluded them for so long.

Two halves of a narrative, tied together by a shared antagonist, these are the parts of Lindbergh's *Listen! The Wind*. Although we spend more time grounded with her than we do in the air, she tells us much about aviation, both the technical, experiential, practical side of it and its human, emotional, and metaphorical facet. She tells us about aviation, not for its own sake, but for what it has taught her about life and living. Unlike the other authors, Lindbergh's life does not revolve around aviation, even in these early days of the 1930s when it consumed so much of her time and effort. Aviation, to her, serves more as a vehicle, a literal vehicle, yes assuredly. But it is a philosophical vehicle as well that lets her study the human condition under extreme circumstances, sometimes exhilarating, sometimes frustrating, sometimes filled with despair. Aviation served another purpose for her as well.

In her "Introduction" to her diaries and letters published in 1980 as *War Within and War Without*, Lindbergh found in retrospect the words that describe the special gift aviation gave her: "Life in the air was beautiful, limitless, and free—if often hazardous—but life on the ground married to a public hero was a full-cry race between hunter and hunted. We were the quarry. We were unable to lead our private lives without being hounded on most occasions by reporters, photographers, and celebrity seekers" (xv). In the air, they could escape the demands of their public life; in the air, she and Charles could be just two people alone with themselves and their lives. Perhaps she knew this from the very beginning, for in the opening chapters of *North to the Orient*, Anne Lindbergh turns to us and says, wistfully we think, for we have read the newspapers for the ensuing almost sixty years:

> I turned to look at the plane. . . .I thought of the two of us, ready to go in it anywhere, and I had a sense of our self-contained insularity. Islands feel like this, I am sure, and walled cities, and sometimes men (40).

10

Louise Thaden: All in a Day's Work

> *There is a decided prejudice on the part of the general public against being piloted by a woman, and as great an aversion, partially because of this, by executives of those companies whose activities require employing pilots.* (151).

> *If you have flown, perhaps you can understand the love a pilot develops for flight. It is much the same emotion a man feels for a woman, or a wife for her husband.* (Foreword).

Louise Thaden was a practical woman and a dreamer all in one, but she could not abide the kind of mythology that surrounded aviation with that aura of the mysterious and the extraordinary. Never one to mince words, she attacks this attitude with the very first sentence of her book, *High, Wide and Frightened* (1938): "A pilot who says he has never been frightened in an airplane is, I'm afraid, lying" (Foreword). This passage sets the tone of her book, a no-nonsense account of the realities of aviation as she saw them. In some ways her realities sound new to us. After all, her life differed in dramatic ways from those of most of the other authors we've explored. Also, her kind of flying differed in significant ways from that of Earhart or Lindbergh, for, although she held numerous records, Thaden's feats dealt more with altitude and endurance, as well as speed in competitive air races, than with the long distance flying of the other two. Hence, the title of her book, alluding to the shape of her kind of flying and, on another level, the shape of her vision.

But beyond the title, Thaden's book offers fresh insights, for she focuses on a facet of aviation relatively untouched by our authors. More than any of our other authors, Thaden focuses on the emotional landscape of flying, not in a self-indulgent way, but as a reminder to her readers that, as with all human beings, pilots must be able to negotiate the unknown and sometimes dangerous territory of their visceral responses to the exigencies of flying. Flying is, for Thaden, too rich an experience to treat any other way. Thus, her book virtually ignores chronology *per se* as a substantive element of structure, although certainly time frames have a place in her writing. Thaden, however, relies more heavily on a thematic structure, tying the parts of her narrative together by related events and the responses to them she and others who were involved expressed. Time is important to her, not as chronology, but as it reveals transformations of feeling and insight.

Thus, for example, what is important about her decision to adopt aviation as a career is not so much what she did, or her parents did, but what they each felt and what she later realizes about that moment:

> It seemed such a long time ago I had said goodbye to my family at Wichita, as I sallied forth in new riding pants, boots, leather jacket, helmet and goggles in approved aviator fashion, for distant California. I had clambered gaily into the front cockpit of the airplane which was to wing me westward toward Opportunity and Aviation. The callousness of twenty-one made it difficult for me to understand the heart-broken tears of a family who felt they were saying goodbye forever. I was too thrilled to feel sad, for my greatest ambition was coming true: I was going to learn to fly! (13).

In this passage, much as in archeological strata, she layers emotions, each slightly different than the one previously felt. Her own first thrill in retrospect seems callous to her; at the same time her family's broken hearts in retrospect suggest over reaction. Both reactions, so normal and right at the time, have, by the time she pens this passage, taken on a nostalgic impersonality. She separates herself from the egocentrism of her earlier reaction, chalking it up to the enthusiasm of youth, without for a moment denigrating the cause of that reaction. She *was* going to learn to fly, and it *was* the greatest ambition of her life. But Thaden adds another twist. That passage refers to a leave-taking barely a year old: "Thinking of that early spring day in Wichita a year before, my fingers gripped the stick more firmly and I breathed deeply of the clean cold air" (13). With the addition of the time element, Thaden provides the reader with a narrative context for her insight: She places herself as she thinks these thoughts in the cockpit of her plane as she seeks the world's endurance flight record, a time we recognize as 1928, fully a decade before she actually penned these words.

In one seemingly simple passage, she has layered four distinct times, identified by the changing emotional response she records: The past (1927) as she left home to learn to fly; the past (1928) of her realizations, which occurred as she flew the long hours toward her first endurance record; the fictionalized subjective reconstruction of the moments of her endurance record (1928 recreated in 1938), with full descriptions of such realities as her cramped muscles and grainy eyelids; and the historical present (1938) in which she writes more objectively of them. This passage, furthermore, illustrates in abbreviated form the structural device that shapes *High, Wide and Frightened.* In her prose these times mingle comfortably, whether in an account of a specific incident or in discussing an entire range of them. Within each chapter she transports her reader from one time frame to another, frequently melding several different ones under the general heading of her chapter themes.

Thus, for example, in her chapter "Engines Quit Cold," she recounts several experiences of near disaster, each conveying something more than its companions about the emotional reality to the pilot of "dead stick landings" or coughing engines. Some of these experiences, traumatic at the time, left only minor reminders that Thaden can endow with a touch of dark humor:

> That's the last time an engine let me down. Although sometimes it's as bad, thinking one is going to quit any moment. No matter whether a failure is anticipated or how careful the preparations made to combat it, when that always dread moment comes, you find yourself resisting

> like a little boy who discovers himself wide awake on the operating table staring in horrified fascination at a monstrous surgeon preparing to disembowel him (47-48).

One event, however, left deep and long-lived scars, and Thaden uses it to remind the reader of the reality pilots must face. She recounts a party at which a young pilot, a favorite among his colleagues, drank too much to safely fly home, so she volunteered to pilot him and his plane back to Oakland, a relatively short flight. Virtually immediately after take off, the engine stalled, and the plane crashed. Thaden was hospitalized, and of her passenger she tells us, "Sandy died that night from a brain concussion. Willie, who had soloed me, told me the next day. Poor guy, I guess no one else had the courage. Willie said he knew how I felt. Only he didn't. There was no feeling, just a numb confused paralysis. But Willie's sympathy was too much and soon hot, salty tears streamed down *inside me*" (emphasis added, 40). Through memory which sounds as vivid in her retelling as it must have felt in the experience, Thaden captures the melange of emotions that make up the entity we call, for want of a more expressive word, grief.

But memory for Thaden is never simple or one dimensional. She recognizes the subjectivity of experience, all experience, and she takes her reader into that realization as well. She continues her account of Sandy's death by telling us, "Outside the sun was shining. Outside it was another day. It was an impossibility to think coherently, to piece things together. The harder I tried to think the more difficult it became to close the gap between dreams and actuality" (40). Impersonal reality and subjective reality diverged, but Thaden draws them together again, in perhaps the only way they can ever be joined. She does not let the clock stop; she does not leave the accident or her emotions to stand unexplored. She recovers to go back to flying because, she tells us, "Airplanes had to be sold, students must be soloed, the books had to balance" (41). On the surface her life continued much as it had before, but from the perspective of her writing a deeper truth emerges:

> There came the realization that nothing I could do would bring Sandy back. Yet there was still remorse. . . . Bitterly, I knew this might never have happened had I not made the error in judgment of attempting a turn [after an engine

> stalled at low altitude]. . . . Two things I blamed myself for: over-confidence and inexperience. Learning this bitter lesson so early, at so great a cost may have left unborn greater catastrophe. It has never been forgotten (41).

Thus, an event from a time remembered becomes a staple of time as now lived.

Thaden's writing is not about time, however, or even time as it changes things. It is about memory and the constant interweaving of past and present memory creates within one's experience. Indeed, the concepts of past and present lose the concrete edges of their definitions in Thaden's writing, sometimes blending, sometimes serving as counterpoints for each other. Past and present each have several facets co-existing at once in her prose. The resulting narrative, then, contains a multitude of perspectives.

The interplay of these perspectives gives texture to Thaden's prose. Significantly, however, sometimes realization of the complexity of her thought comes to the reader only in the memory of what she has written in previous portions of the book. Nothing illustrates this any better, perhaps, than a passage in which Thaden seems to re-evaluate one of her most firmly held opinions about aviation: "It occurred to me I might have been wrong in my judgment of pilots; perhaps after all, they are a race apart because of the gamut of emotion they may experience in the course of an hour's flight" (137). Virtually every account she has provided has affirmed that wealth of emotions; virtually every account has detailed some form of crisis or near crisis. Legitimately, one may ask what manner of person would consistently engage in such a career. Thaden's answer is that the professional aviator must be a person who can live comfortably within a paradox, for that is the way she expresses the nature of aviation. On one hand, she realizes, "flying, especially concentrated flying, saps vitality. It takes an important part of your life away. A whole lifetime, two, three or four lifetimes, are sometimes crowded into a few short hours" (137). That vitality, sapped away, cannot be used for family, for living. On the other hand, however, she knows as well, "When you haven't been in the air for a while, it becomes madness—the desire to fly. Perhaps because flying is the only real freedom we are privileged to possess" (138). Aviation to her, as she affirms in countless passages within the book, represents peacefulness, serenity, freedom, the chance to think. Perhaps that contradiction in need, demonstrated so clearly in her own life, *does* mark a people apart from the ordinary.

Certainly, Thaden sees this paradox as complicating the life of the aviator, especially the woman aviator. Remarkably introspective, she gives voice to the conflict she sees inherent in her own life—the fiercely intense love she feels for her family and the passionate craving she has for aviation.

Thaden's book contains many references to husband Herb, son Bill, and later, daughter Patsy. No question arises that Thaden's family held a central part in her life. Along with Anne Morrow Lindbergh among the aviators, Thaden seems to have lived as close to a typical woman's life as is possible, given her love of flying. Throughout her narrative, we see vignettes of Thaden with baby Bill in tow visiting the airports, we see her working beside Herb in building the aviation business that sustained them, we see her engaged in the negotiated truces and the moments of unity that accompany any marriage, we hear her tell of the two children wanting exciting transportation such as automobiles and trains instead of the commonplace airplanes. We see her, in short, living her personal family life, not just as a daughter to worrying parents but as a parent and spouse engaged in the daily living of life. We hear her as well admonishing us, "Rearing children is the one real accomplishment of woman. In comparison all things else pale. Fame is unstable, fleeting. Our children, and their children shall be my monument forever" (95).

Even so, for Thaden such valuing of children and of her role as wife and mother did not exist as an exclusive value, for she follows the above statement, made in that subjective past she uses so often—in this case to suggest her experiences as she awaited the birth of her first child—with the admission, "Summer passed by while I fought oppression" (95). Again, we are brought close to her conflicting passions, for just moments before, we had heard her admit, "As much as anything else, I missed the soothing splendor of flight. . . .To feel, if only for one brief moment, that I could be master of my fate—that is what I missed!" (94-95). In her life, more than is expressed in the lives of our other aviators, these bonds of loyalty and kinship in its broadest meaning frequently collide. Her life, as she describes it, consisted of periods of isolation from one of her loves or the other. She sees in her life the dilemma of the woman pilot:

> To a psychoanalyst, a woman pilot, particularly a married one with children, must prove an interesting as well as an inexhaustible subject. Torn between two loves, emotionally confused, the desire to fly an incurable disease eating out your life in the slow torture of frustration—she cannot be a simple, natural personality (139).

Yet, Thaden does not portray her life as one of isolation or loss. Instead, she provides in this book a picture of balance, focusing most often on aviation—for that is, after all, her ostensible subject—but also permitting glimpses at the more personal side of her life, as well.

The professional side of Thaden's life inextricably deals with gender, for more than many of our other women aviators, she expresses deep frustration with the male gender, whether they be pilots or executives or passengers. Their difficulty adjusting to women in their all-male domain took several forms, most of which Thaden experienced first hand. She recounts, for example, being told, when taking the flying portion of her examination for a commercial pilots license, that since she is a woman, the examiner will be more demanding of her than he would a male pilot because he doesn't want to be blamed if a woman he certified should have an accident. On numerous occasions, she also experienced the patronizing attitude confronting women aviators who strive to succeed in a predominantly male profession. The prestigious Bendix Race illustrates the extent of the patronizing quite well. In 1936, the first year women pilots could compete in the formerly all-male air race, the founder of the race offered a $2500 award, "for the female pilot who finishes first regardless of her position in the race itself" (167). To the organizers of the race, a woman's winning was nothing less than inconceivable. In their eyes, their offer was a magnanimous gesture of consolation to show women they were welcome, even if not competitive contenders for first place.

As she portrays the decision, Thaden as pilot and Blanche Noyes as navigator entered the race virtually on a whim, albeit a whim backed by three weeks of intense preparation of both plane and themselves. Contrary to all advice from male colleagues urging her to "Open this damn thing up," she flew her own race, steady and careful not to overtax the engines:

> I learned long ago from hard-earned experience gained in several long cross-country races that a race is not always won by the fastest plane; that good common sense in taking care of engine and equipment sometimes proves the winning factor. Yet I knew too that luck plays the leading role in such dramas of the air (176).

History provides the result of that famous race. Thaden won first place, taking the prize in the first race of its kind open to both men and women. "Mr. Bendix and Cliff Henderson were there—a little disappointed, I think, at the apparent outcome of the race. They looked so crestfallen" (182). She tells us, too, that by the time the evening newspapers were released, the contest committee "had changed the name of Mr. Bendix's extra $2500 to the first woman completing the race from a'Consolation Prize' to a'Special Award'" (183).

As aircraft demonstrator, Thaden also had ample opportunity to run into gender-based problems, originating from a combination of the male ego and the prevailing prejudice against women in aviation. "There must be some psychological reaction which reacts on the male in the form of an urge not only to show the female the prowess of the male, but in the process, turning the airplane wrong side out and hind side front" (187-188). She describes herself in her role as company representative, whose job it is to sell airplanes, being torn between politeness and the urge to rip the controls from an incompetent male pilot's hands. But the male reaction proves to be only half the equation. Part of her dissonance springs from her awareness of societal expectations of gender roles, of society's expectation that she as woman should know her place. Such realization on Thaden's part stymies her natural responses: "It was during this demonstration trip that I came to the conclusion that women pilots have an inferiority complex themselves. At least this woman pilot. I dislike taking the controls away from a man" (188). In a later passage, she returns to this problem of taking charge, admitting that part of the problem is her awareness that for her to do so would embarrass the man. Part of the problem, also, comes from an entirely different source: "It isn't good business to invoke a potential customer's wrath" (193). Again, as readers we realize that she finds herself in a bind remarkably familiar to us, for even if she overcomes her hesitancy to challenge the ego of the male customer, what will happen to her job? The question holds particular pertinence when one remembers that these were years of the Depression, and jobs, especially jobs for women aviators, were scarce. Although Thaden herself does not voice this question in her text, we as readers must ask it for her, for such was the reality under which she flew.

That reality did not keep Thaden from flying or from winning awards and setting records, nevertheless. One of the most prominent features of her accounts of those accomplishments, however, is the camaraderie she describes among the women pilots around her. More than any of our other writers, she populates her narrative with the interactions of a group of women, well known to each other and the public, sharing a passion for aviation and a joy of life, competing with and against each other, working together on common goals, all in a spirit of family. While Thaden tells of her own experiences, she also provides insight into the world of aviation, especially for women pilots, in the 1930s. Its needs and its expectations differ dramatically from those of today; indeed, she describes a much simpler form of aviation than the one which has ultimately developed. Nevertheless, Thaden spent the middle years of the 1930s working with the WPA and the Bureau of Air Commerce on a project originated by Phoebe Omlie and Amelia Earhart to hire women pilots to travel around the country encouraging towns to provide aerial markers for pilots, simply a matter of painting the town's name on barn roofs or other appropriately visible places. As she reminds us, commercial advertisers had been doing so for years, so the idea had precedent and logic behind it. During her years with this service, she worked closely with women pilots, and those pilots, as well as others, occupy comfortable spaces within her book.

Thaden gives Blanche Noyes, her navigator during the Bendix Race of 1936, quite a bit of space in her story. Combining both the personal friendship and the professional aviation facets of their relationship, Thaden tells of the excitement they shared during the early morning hours before the race began. Some of it dealt with practical matters, for example deciding who should bail out first in case of emergency, each concerned about the welfare of the other. Some of their excitement sprang purely from the emotions, the thrill of competition. Thaden's portrayal of the flight reveals the confidence with which she acted on Noyes's calculations and, of course, the justification of that trust. It shows, also, a playfulness, for Thaden and

Noyes bet on their arrival time at the finish line, Thaden choosing 5:11 PM and Noyes, 5:08 PM. They crossed the finish line from the wrong direction—"'Nothing like coming in through the back door!' Blanche yelled with a grin" (181)—at somewhere between 5:09 and 5:10. They called the bet a draw. Thaden's description of the flight, officially recorded under Thaden's name alone, shares the credit with Noyes. It stands, in the chapter she devotes to it, as a two-woman victory.

She also gives special space within her narrative to Frances Marsalis, with whom she set the refueling endurance record in 1932. What emerges in Thaden's account of her refueling endurance flight with Frances Marsalis is a chronicle of two women working successfully together under far less than ideal conditions, finding time to share laughter as well as the tedium and monotony of flying endless circles around the airport. She portrays a real working relationship between two friends. Given the restrictions of the situation—more than five days aloft in a cramped airplane with exhaust fumes and constant roaring noise, going nowhere except those endless circles, dependent on ground crews and the refueling contact for food and fuel—such a flight offers little incentive except the record. Thaden's and Marsalis's first attempt ended after eighteen hours when, flying too close to the refueling plane, Thaden sheared a wing tip off the plane. Having to land for repairs invalidated those first eighteen hours, indeed erased them for all practical purposes, even though it allowed the two aviators to reassess the equipment they took with them and catch up on needed sleep. It allowed them, also, a new and not too pleasant perspective on their task:

> At four o'clock on the 14th of August, we took to the air again, each wanting more than anything to be doing something else, yet afraid to quit for fear "they" would think us yellow. Eighteen hours had convinced us endurance flying was *not* the fun we had anticipated (115).

Thaden describes the rigors of the ultimately successful flight, beginning with their flying in three-hour shifts which by the end of the flight had turned into one-hour shifts, and the immense problems they had staying awake and alert during the long days and nights of the flight. Comfort was not a concern—or a possibility. Sleep itself offered no real solace, for as she tells us, they experienced "Fitful sleep filled with weird dreams of head-on collisions, of tail spins," but she follows that lament immediately with details of the parts of the flight that made it bearable: "Wondrous dawns, with the sun peeping cautiously over a far distant horizon, rising in gallant heroism from the cold waters of the Atlantic" (119).

With some chagrin, Thaden also tells of giving in to pressure from the organizer of the flight who, concerned that the monotony of the flight would affect its publicity potential, and with the compliance of the reporters covering the story, urged the women to fake an appendicitis attack for Frances. With some amusement, for they were bored, too, Thaden and Marsalis agreed: "'Oh well,' we shrugged, 'why not?'" (120). After the fact, Thaden recounts her thoughts about the deception, barely covering the amusement she still felt, but providing sufficient reason for repenting her actions:

> I don't know why anyone thought I could stay in the pilot's seat and take care of Frances in the rear of the cabin, nor how I could fly for 16 hours without relief. Anyhow, Frances was much better the following day, due no doubt to my diligent application of ice packs, so we winged steadily aloft, the crisis over. Never again will I be coerced into duping the reading public, for it was just a week from that day exactly that I really suffered a severe attack of appendicitis (121).

That same, somewhat mischievous, sense of humor followed Thaden in her relationship to Marsalis. In a later chapter, she tells of becoming lost flying through the mountains of Pennsylvania, and after landing at a small airport to discover she is 200 miles off course. She recounts the consolation she receives from the field manager; she shouldn't feel too bad, he says, because "Even *good* pilots get lost in these hills, all the time" (140). In response to his following request for her name, she responds, "Frances Marsalis," even signing the pilot's register in her name. The mark of friendship shows, however, as Thaden's following actions indicate: "Although I was ashamed, at the same time the story was too good to keep; I had to tell it, to Frances first of all" (141). She knew Marsalis would understand and would laugh with her.

The bond that grew between the two aviators, solidified in their successful endurance record flight and extending into other facets of their professional life, united them in their love of aviation. Thus, when Marsalis was killed in an air race some time later, Thaden's response contained that mixture of pain and understanding that mark the aviator's approach to death:

> I could not help thinking that although Frances may have missed a few pleasures—perhaps a little additional fame, too—that she could never have been as happy here as she must be in the Valhalla of flyers. For flying and things that fly were her life,—her meat and drink and sleep,—and the flying game is a tough, a hard nut to crack, . . (147).

That attitude toward the danger always facing pilots, especially those who chose to dare time and distance and altitude as did Thaden and her contemporaries, permeates the Postscript, "A.E.," which she attaches to her autobiography, for Thaden cannot end her book without notice of Amelia Earhart whose disappearance still retained in 1938 its aura of shock. Describing herself as a neophyte, Thaden admits that Earhart served as a role model for her: "Dignified, reserved, yet natural, and, above all, human in her accomplishment" (257). She recounts several episodes in their developing friendship which spoke to a goal they shared: greater recognition and opportunity for women in aviation. Both of them chaffed at the patronizing inflicted on women aviators by their male counterparts. Both of them had been involved for years in creating and sustaining opportunities for women to become more accepted professionally. She recalls trying gently to dissuade Earhart from attempting that last, risky around-the-world flight, only to be reminded by Earhart that as the woman who won the Bendix in a plane with an oversized gas tank partially blocking the escape route, "'*You* can't talk to *me* about taking chances!'" (262). In fact, she tells us, Earhart told her that, in case of disaster, Thaden should send water lilies as appropriate flowers for her memorial.

But Thaden can't leave the matter there, and she creates a much finer memorial to the woman and aviator:

> Perhaps it is because I have known Amelia for so long that I find it difficult to draw a word picture of her. Perhaps that is why it is impossible adequately to describe her staunch fineness, her clear-eyed honesty, her unbiased fairness, the undefeated spirit, the calm resourcefulness, her splendid mentality, the nervous reserve which has carried her through exhausting flights and more exhausting lecture tours. . . .
>
> She will live. So I have said no farewell to her. As she invariably ended letters to me, so I say to her, "Cheerio!" (263).

And thus, she closes her book, at once an autobiography of her own life and a testimony to the women who lived and died for aviation, who shared with her mutual dreams and experiences. Fittingly, for it captures the spirit with which Thaden shares these experiences, the last word of her book belongs to both a treasured friend and colleague and to her own world view.

Thaden's life encompassed enormous complexities of emotion and circumstance which caused her to explore the far reaches of her own dedication to life. Perhaps because of her willingness to engage in that exploration in the pages of her narrative, we come to a deeper understanding of what she found in aviation, and what she could not find there. Her passion for flying, for friendship, for family, and for living resounds throughout her story, a story which she would remind us is of an ordinary person facing extraordinary circumstances as an ordinary day's work.

11

Sisters of the Wind

The May 12, 1928, issue of *Literary Digest* carried an article which proclaimed that aviators speak a language of their own. While it listed some terms that had migrated into everyday use, the article's emphasis was placed clearly on those bits of language that had retained the odd, occupational orientation of the aviator. In many ways, this article published in a decidedly nontechnological journal serves as but one manifestation of the mythology that surrounded aviation in the 1920s and 1930s.

While the myth usually found expression in predominantly male terms—"Wizards of the Air," "Apollos of the Sky," "Sons of Dedalus," which in their very wording excluded women and which we have difficulty translating into female analogs—the women were participants also. In telling their stories, they of necessity had to respond to the myth, to write around it or through it or beyond it in order to make themselves heard. As we read their stories, we find them weaving new mythologies from the threads of the old, sometimes as a Penelope who used her weaving to control her own destiny and sometimes as the Grandmother Spider of Native American narratives who used her weaving to open up new worlds of experiences for herself and others. The task for these women aviators was complex, for it asked them to deal with two forms of myth: The one that portrayed aviation as requiring "male" strength and courage, and the more dangerous one that portrayed women as fragile, emotionally unstable, and technologically incompetent.

They had basically four possible responses: They could choose to adapt the existing mythology to their own uses, circumventing it by continuing to tell the story of their own accomplishments within the framework that would seem to deny them; they could portray themselves within the male mythology, giving themselves male roles; they could demythologize aviation and deflate the negative female stereotype; or they could ignore the myth altogether and create a new framework for their accomplishments. Our authors illustrate all these options.

The first choice, playing to female frailty while at the same time circumventing it, appealed to several of the aviators, especially those writing the earlier autobiographies written as the 1920s turned into the 1930s. So we see Mary Bruce deciding an around-the-world flight on the vagaries of a shopping spree. The outrageousness of her portrayal of that fateful shop window introduces us to the female stereotype: "Just like a woman," she succumbs to impulse, seemingly oblivious to the rational judgment that such a journey was beyond her capability—she couldn't even fly a plane yet. Nevertheless, even as she conforms to the stereotype, as she does periodically throughout the narrative, she adapts it to her own purpose, showing her determination to make the journey, her competence in acquiring the necessary skills, her courage and creativity in meeting unexpected hardships, and her achievement in completing the journey on her own terms—always on her own terms. Lady Heath, as well, even as she gossips and flounces from one outpost to another, enjoys her role as an unpredictable woman intruding on the stolid male domain. With her evening gowns and tennis gear, it seems to please her that no one seems to know where she'll pop up next. Underneath it all, however, Lady Heath portrays herself as a woman in charge of her own destiny, obeying the rules when she chooses and ignoring them when she doesn't. In this she's much like Violette De Sibour who enjoys the reputation of "madcap aristocrat" and denies that she's much of an aviator, even as she takes charge of an around-the-world flight. All of these women adopt the pose they felt society expected of them. But all of them also managed to use that stereotype as entree into their records of extraordinary achievement. Perhaps they thought the times demanded circumvention; perhaps they merely recorded the self-contradictory truth of the lives they led.

Others of the aviators chose more direct approaches to their stories. Josephine Bacon, for example, identifies herself with the maleness of aviation. She eschews any trace of the female, which she distrusts. Amy Johnson, too, for other reasons glories in the mythology of aviation and the special places of aviators, hating to see the era end. She recounts her story from her status as pilot, refusing to acknowledge any subsidiary role in the pantheon, even as she decries the lack of public recognition women are accorded. But Jean Batten takes the more active strategy of actually identifying with the explorer myth, comparing herself to Columbus and marking the records she set, the records she has taken from men. The tone these women adopt is less deferential, more assertive. They expect credit where they believe it to be due.

But the majority of the women chose to attack the mythology altogether, both of its parts, and they demonstrated no reticence in doing so. As passenger, Harriet Camac accomplishes this by showing her readers that flying can be an efficient option for travel, even for a woman. Through her descriptions of airplane cabin design and landing field accommodations, she makes the trappings of flight more visual, more familiar, and therefore less mysterious to her audience. Along those same lines, Pauline Gower, Amelia Earhart, Louise Thaden, and Lady Bailey all speak in matter-of-fact tones of the routine and often mundane elements of flying. These women describe piloting in terms of everyday hard work and training, and of the sometimes boring experience that gives one the flexibility to avoid disaster. They also discuss in no uncertain terms the need for women to develop the technological and mechanical skills necessary for careers in aviation. Through their own lives and in their work with other women, they stress that society must abandon its myopic reliance on outdated expectations for women. Such a change in attitude becomes a virtual crusade for them.

Two of our women, however, chose to ignore the mythology altogether and explain aviation in completely different terms. As passenger, Marie Beale saw aviation as a literal vehicle into a historical past, which she described in metaphorical terms. Aviation linked for her the past and the present and made a form of intellectual time travel possible for anyone—male or female—willing to try it. She openly encouraged her readers to explore the past through the technology of the present, but the technology always takes second place to the real wonders of the historical past. Anne

Lindbergh, too, saw the metaphorical dimensions of aviation. She used it literally, but more importantly metaphorically, to describe the journey into human nature it allowed her to take. She writes of aviation not so much for its own sake, but for the flights of the mind it encourages its aficionados, armchair or not, to take.

In responding to the myths that surrounded aviation, however, they all addressed, in one way or another, the crippling effects of society's stereotypical view of women. The effects of that stereotype permeated all elements of their lives, as the following example—which at first glance seems trivial to us—illustrates. Virtually all of these authors at one time or another in their autobiographical volumes mention the matter of their clothing. Bacon laughs at the strictures of the past; Heath, Bruce, DeSibour, and Batten discuss the need for evening dresses in their luggage; Lindbergh explains the need to be frugal with clothing because it adds unnecessary weight to the airplane; Earhart discusses why she dresses in trousers and leather jacket and why they are preferred in airplane hangars; Murray records that at an official luncheon during their trans-African flights, Lady Heath looked splendid in an evening gown, while Lady Bailey wore her flight clothes; the passengers Beale and Camac reassure their readers that one can fly in everyday clothing. The list goes on. Why, we ask, do they spend so much time and attention on clothing? Before we succumb to the temptation of saying "Just like a woman to be concerned with fashion at a time like this," it behooves us to look once again at the stereotypes that confronted them and the pressures they felt from both society and the media. Everywhere they turned, society and the media expected them to retain their feminine roles even if they did dare to enter the masculine world of aviation. The media reported the styles and colors they wore, the public expected them to remain ladylike. Headlines followed them with accounts of windblown hair (Earhart faced chronic media disapproval of her hair, once being urged by a columnist to comb it before stepping from the plane) or appropriate makeup (Mary Bruce's landing in Japan—a record-setting flight—was reported first with notice that she applied lipstick before she left the plane). Unconventional clothing, such as trousers, had to be justified; the touch of glamour, such as a fresh frock or an elegant evening gown at the flight's end was much to be desired. Woe unto them if they failed to meet public standards for dress and grooming. All of the women felt the sting of societal focus on what was to them, and to us, an irrelevancy. Nevertheless, this irrelevancy, too, became a part of the mythology.

While they address the issue in different ways, the women who wrote of their experiences show a determination to dispel the negative force of the mythologies surrounding aviation—the mythologies that confronted them every day— by replacing them with those of their own design. Through their language patterns, imagery, allusions, personae, and other rhetorical devices they define their own perspectives of themselves as aviators and of women's place in a world in which aviation seems the norm. Their individual autobiographies, both separately and collectively, reveal the emergence of a self-conscious feminism on their parts, one that they adopt not in order to accomplish their goals, but to explain them to a society unresponsive to having its cherished beliefs and order challenged.

Virtually all the women aviators also deal with the women's political/social issues of their day, and as one reads them chronologically, one realizes that they recreate in their particular stories about their aviation experiences the shifting pattern of women's overall relationship to society during the 1920s and 1930s. Thus, the accounts written in the 1920s, for example by Amelia Earhart or Lady Heath, focus on the promise of aviation and on the spiritual liberation available to those women who learn to fly. In those heady years after the success of the Women's Suffrage Movement both in this country and in Britain, the skies literally seemed to be open to women of courage. In deference to the developing changes in women's perspectives of themselves, many of these texts also encourage women who do not want to become pilots to learn more about aviation and aerodynamic theory, to learn sound business practices and endorse them for aviation, and to demand more scientific and technical training for females in the public school systems of the nation.

As the 1930s developed and women aviators became public celebrities, their agenda and their tone changed. The reluctance of the aviation industry, the government, and the public in general to foresee a legitimate, practical place for women in cockpits (again, a telling word), board rooms, and aeronautical design studios drove the women aviators to express a growing frustration with the system. Even gentle Anne Morrow Lindbergh expressed irritation with the press which asked technical questions of Charles and turned to her with questions about how she kept the plane comfortable and their food fresh on long trips. Many of the texts, notably Louise Thaden's, Amelia Earhart's, Amy Johnson's, and Pauline Gower's, contain passages demonstrating that aviation is not gender based, that what matters is knowledge, experience, and a reliable airplane. While these texts still rely on ironic humor to attack the absurdity of prejudice against women in aviation, the humor surrounding these issues shows a sharper edge than the gentle prodding of the earlier decade. Indeed, by the end of the 1930s, Thaden rails in disbelief at a government preparing for war, desperate for young men to train as pilots, completely ignoring and ultimately rejecting the licensed, trained women pilots volunteering for service to their country.

Ultimately, however, as we look at their various accounts, we must ask ourselves the central question: What do we learn from these accounts? What do they tell us about women's lives, even those lived in similar circumstances and similar times? For one thing, the urge to tell our stories is a central part of our heritage as women. Refusing to believe the myths that belie our collective experiences, we find in our past women who have recorded their stories with voices we once thought muted. In their narratives, we find not the silences of women robbed of their speech, but the shouts of women who have used the vocabularies of their day to tell the truth of their lives. If we find convolutions in their narratives, those convolutions reflect the necessities of women's daily existence. This, too, is necessary for us to know. The narratives speak on several levels, and by their very natures show us a complete truth.

Individually, we find women of great courage and a sustaining humor who challenged their society and succeeded in proving it wrong. Today, we can only look with awe at the strength they showed, the obstacles they overcame. Collectively, we find a solidarity and sisterhood among women of similar times and circumstances, a comfortable companionship not often acknowledged by the official records of women's achievements. This discovery of simple friendship among like-minded women proves valuable to us, for it shows a health that permeates women's relationships with each other. It provides reassuring evidence: We have not been crushed by the mythology; our foremothers have been too strong, and they have left us a powerful legacy.

But we learn something, too, far less positive, for we must admit the fragility of the traces women have left behind. Some of these aviators verge on the totally forgotten, their names submerged by other voices, other times. Perhaps, however, this knowledge, too, empowers us, for it prompts us to reclaim that powerful legacy and to remember our heritage—remember to hear the stories.

Epilogue: What Became of Our Aviators?

One of the peculiarities of autobiographical writing is that it never tells the end of the story. Always, pages or even chapters remain to be written, and such is surely the case for the women in this volume. Some of them lived long and prospered, some did not. But closure demands that we find an accounting of those stories. We have been involved with these women, heard them speak to us, and come to see them as individuals with a wealth of experiences and an urge to communicate about them. We feel a need to know their fates.

The search for that information seems itself an expedition of sorts, for finding these women in official ledgers has not always been a simple task. Some of their lives, of course, have been well documented. Amelia Earhart, Anne Morrow Lindbergh, Amy Johnson, and Louise Thaden have all been the subjects of scholarly biographies; Earhart and Lindbergh have sustained heights of public attention over the years. Indeed, they have been the subjects of hysterical journalism from various decades, Lindbergh finally finding a peace with the media still denied to Earhart. Others have virtually disappeared from the public eye, for example Harriet Camac and Stella Wolfe Murray, who have proved as elusive as mist. Still others, such as Marie Beale, have been remembered for parts of their lives very far removed from their own narratives.

Of course, no one lives an entire life without touching other people, and traces of that life remain in official records, business transactions, and personal memories of family, friends, acquaintances. They exist in the artifacts that become the biographer's tool. Several of our authors have been the subjects of biographers' searches, which produced thorough and credible studies—for example, Constance Babbington Smith's study of Amy Johnson, Ian Mackersey's of Jean Batten, Mary Lovell's and Doris Rich's of Amelia Earhart, and Dorothy Herrmann's of Anne Morrow Lindbergh, to name the most recent. Ian Mackersey's research on Jean Batten's life shows us some of the biographer's skill in action, for his book weaves his own process of research into his discussion of Batten's life and death. He chronicles the time, expense, and frustrations that accompany the recreation of a subject's life. Such a process is far beyond the scope of *Sisters of the Wind*, for our purpose seeks not the recreation of our subjects' lives, but the exploration of the images of themselves they wanted to convey.

Nevertheless, natural curiosity causes us to want to know what happened to these women who wrote to us so confidently of their achievements. For this kind of biographical information, a record of their public impact in its broadest outlines, we turn to the media that reported their accomplishments so well. What did they record after the flights were over? What do they leave us of those inevitable traces of lives once so famous and newsworthy? What follows is the result of a careful study of major public sources, such as the New York *Times*, the London *Times*, the Washington *Post*, various magazines, biographical encyclopedias, indices, obituaries, and the like.

In recounting what remains of their accessible public record, I am also reporting on the way we as a society remember women's lives, on the way we preserve their memories, on the way we keep track of their value. Thus, some of these accounts may seem somewhat slight, for little was written about some of these women beyond what they told us in their own narratives. Some leave large gaps of time unaccounted for, for some of the women retired from public notice of any kind. Other accounts may seem somewhat strange, especially when compared to the perceptions of themselves these women portrayed in their own writing. Nevertheless, to the best of our collective memories, these are the continuations of our aviators' stories:

Bacon, Gertrude (1874-1949): Bacon never learned to pilot a plane herself, but she continued to write about aviation and to establish a series of "first woman to" records, including the first woman to fly as passenger in a plane looping the loop. Legitimately, she was the first woman in England to ride in an airplane. She was also an authority on wild flowers and an amateur astronomer. During World War I, she joined the Red Cross. In her fifties, she married T. J. Foggitt in 1929. She wrote numerous articles and books, including a lengthy biography of her father, always under her own name,

Gertrude Bacon.

Bailey, Lady Mary (1890-1960): Most accounts of Mary Bailey's life begin by remarking that by the time she began flying in 1926, she was already the mother of five children, having married South African millionaire Abe Bailey in 1911. She began her aviation career by immediately setting a record as the first woman to solo across the Irish Sea (1927) and continued it by setting a 1927 altitude record. She is most often remembered for the journey she wrote about, her 1928 round-trip flight from Croydon to Cape Town, returning to Croydon. That journey, as well as her participation in various international aviation competitions and her writings about aviation, contributed to her being named a Dame of the British Empire in 1930. The late 1920s and early 1930s were her most active period as an aviator, but in 1940 she also ferried military planes for the British government. Throughout her life she befriended other aviators, such as Amy Johnson who wrote fondly of her in *Sky Roads of the World* (1939). She died at her home in South Africa in 1960. Of the intervening years, very little public media record remains.

Batten, Jean (1910-1982): A New Zealander, Batten went to England in 1929 to study music, but found flying instead. In 1934, after two unsuccessful attempts, she set the world's record for women's solo flight from England to Australia, overturning the record previously held by Amy Johnson. Because of these activities, Batten became one of the most publicly visible of the British aviators, earning the media's title as the "Garbo of the Air," but also inviting criticism of her commercialization of her own image. Throughout her flying career, Batten, an intensely private woman about her personal life, was criticized for arrogance and egocentrism. While generally conceded to be the better aviator of the two, she was often compared unfavorably to her contemporary, the beloved Amy Johnson. Nevertheless, for both psychological and financial reasons, Batten had to capitalize on the fame her flights generated. She endorsed many products, some relating to aviation and some not, and she engaged in aviation-related writing and broadcasting. Her books received generally scathing to mediocre reviews, some of which seem to be more *ad hominem* attacks than they were critiques of her writing skills. In 1938, she launched a public appearance tour of

England and various parts of Europe where she socialized with some of the rich and noble families. She returned to England in 1939, was rejected for employment with the Air Transport Auxiliary, and spent the wartime years on a fundraising tour of England. After the war, she and her mother lived a self-contained life, first in Jamaica and later on an extended nomadic tour of Europe. Intensely lonely and grief-stricken after her mother's death in 1966, Batten threw herself into a round of public appearances in the 1970s commemorating the era of 1930s long distance flying. In 1982 she settled in Majorca to such a reclusive life that her death in 1982 was not publicly known until 1987, after extensive detective work by family and Ian Mackersey to find out what happened to her. Their detective work led to the discovery that Batten, who had left an estate of approximately £100,000 had been buried in a pauper's grave with 150 other bodies.

Beale, Marie (Oge) (1881-1956): Born in California, Marie Oge married Thruxton Beale in 1903 and is remembered in her obituaries as his widow. In her role as wife of a diplomat and philanthropist, she served for close to fifty years as mistress of Decatur House on Lafayette Square in Washington, D. C., holding formal dinners remarkable for her insistence on using candlelight. In fact, many rooms of the historic house were not wired for electricity even at her death. In 1954, as a correlation to the interest in historical preservation we saw in her books, she bequeathed the mansion to the nation as a national monument. She was decorated in Italy because of her work there in historical preservation, especially in Venice. She died in a Rome hospital (some accounts say Zurich), after having to suddenly abandon a trip to Athens. Some accounts list the cause of death as unknown, some as leukemia. Beyond this information, nothing else is recorded of her in the notices of her death, which, as we note, use more space to discuss Thruxton Beale and Decatur House than the accomplishments or life of Marie Oge Beale.

Bruce, Mary (1895-1990): Although she originally wrote under the name Hon. Mrs. Victor Bruce and was written about under that name, she divorced Hon. Victor in 1941 after fifteen years of marriage, never marrying again. Her flight in 1930-31 clearly gained media attention, for many reasons. Early in the flight, the New York *Times* claimed she was trying to beat Amy Johnson's record time to Australia, while by the time she arrived in Japan, it had turned to more feature-oriented reporting with a headline that focused on her applying make-up before meeting the press (Nov. 21, 1930). Aviation was but one of Mary Bruce's accomplishments, for she also set numerous speed and distance records in motor racing and speedboating during the late 1920s and early 1930s and won awards in equestrian show jumping in the late 1930s. By the mid-1930s she turned to more business-oriented aviation, flying first in an air circus (where she and Pauline Gower and Dorothy Spicer toured together during the summer of 1934) and then establishing her own air transport business which carried both freight and passengers. She also established a small commuter airline linking London with the surrounding towns. During World War II she channeled her aviation business to the war effort, including operating an air ferry service to France and opening a repair shop for military planes. In her later years, she was remembered for test driving Ford cars as late as the 1970s. In her lifetime Mary Bruce wrote six books, including a collection of aviation short fiction and a 1977 autobiography with a title that perhaps captures what she thought of her life, *Nine Lives Plus.*

Camac, Harriet (?-?): Clearly the one of the least written about of the women included in this study, Camac left very little trace in the news media. Brief news accounts of her Imperial Airways flight in 1928 mention that she was prominent in New York society. As a young woman she belonged to the Junior League and was involved in several amateur theatrical presentations. In 1935 she married Dr. John Edward Elmendorf, Jr., who was a noted medical researcher and malaria expert. At the time of his death in 1960, they shared homes in Canada and North Carolina. She survived him.

De Sibour, Violette (?-?): Newspaper accounts of the De Sibours' flight seemed as interested in the Vicomtesse's wardrobe as they were in anything else about her, and many found newsworthy that she had chosen to wear trousers instead of a skirt in the airplane. Accounts of the De Sibours' eventual safe arrival in London at the end of their journey recount that her first words to her father were to ask for a pocket comb and that the three frocks she took with her lasted the entire ten months. We are assured by that same article that she had, however, saved a fresh one for her arrival.

Virtually all accounts also mention that she is the daughter of H. Gordon Selfridge, American merchant in London. The New York *Times* adds to their coverage of the trip that Selfridge was formerly the manager of the Marshall Field store in Chicago and that the engine in their plane was manufactured by the American Moth Company in Lowell, Massachusetts. Violette De Sibour supplemented the news accounts by writing letters to family during the flight and these letters became a useful resource as she wrote *Flying Gypsies*. While shooting in Indochina, they received word of Jacques's father's illness, but before they could change their plans for a return to Paris, they were notified of his death and decided to continue their around-the-world journey. Violette De Sibour met Amelia Earhart after Earhart's 1932 solo flight across the Atlantic Ocean, and the two remained friends until Earhart's death.

Earhart, Amelia (1898-1937): So much has been written about Earhart, and so much speculation exists about her fate, that the mention here needs to be brief. In 1922, Earhart became the first woman to become licensed as a pilot by the F.A.I. During the 1920s she became interested in the business and professional side of aviation and became investor in Dennison Aviation Corporation and president of the Boston chapter of the National Aviation Association. With pilot Ruth Nichols she became involved in increasing the participation of women in all aspects of aviation. After the completion of the Friendship flight in 1928, she became the aviation editor for *Cosmopolitan* magazine, a position she held for almost two years. In the late 1920s, she bought her first airplane from Lady Heath, who inscribed its fuselage with the following: "To Amelia Earhart from Mary Heath. Always think with your stick forward." After his divorce, in 1931 Earhart married George Palmer Putnam, who also became her publisher and was publisher for several other aviators. In 1932 Earhart soloed across the Atlantic, becoming in reality the first woman to fly across the Atlantic. Thereafter, she engaged in a series of long distance flights taking her to many parts of the American continents and the Pacific Ocean. In the mid-1930s she became associated with Purdue University's aviation program, and planned to work more extensively with it after she retired from long distance "stunt" flying, which she planned to do after the 1937 flight. She and George Putnam were in the process of completing a move to the West

Coast when she embarked on her around-the-world flight.

Gower, Pauline (1910-1947): After the death of her mother in 1936, Gower could no longer remain a full-time partner in the air garage business she and Dorothy Spicer founded. The two women, however, remained close friends. In 1937 as a member of the Women's Engineering Society, Gower presented a paper on methods for preventing ice buildup on aircraft. Her interest in women's contribution to the war effort bore fruit when in 1940 she became commander of the Air Transport Auxiliary made up of accomplished women pilots charged with delivering fighter planes from factory to front lines and other necessary flying jobs associated with combat. Amy Johnson ultimately became one of those pilots. As Air Transport Commander, Gower traveled to the United States in 1941 as a consultant for establishing an air ferry service in this country. In 1943 she was appointed to the Board of the British Overseas Airways Corporation, acting as an advisor to the Air Ministry on issues concerning women air passengers. In 1945, she married W. C. Fahie. She died giving birth to twin boys, who survived.

Heath, Lady Sophie (Mary) (1896-1936): Usually considered in the news media of the late 1920s England's most noted—and we might add, flamboyant—woman flyer, Lady Heath wrote extensively about aviation during the 1920s and 1930s. During World War I, she served as a motorcycle dispatcher, and after the war she earned a degree in agriculture from the University of Dublin. She also ran a coffee plantation in Kenya for some years. In 1928, she was the first woman to fly solo from Cape Town to Croydon (the reverse of Lady Bailey's flight), as she wrote in the selection we examined. Earlier, in the 1920s, she had championed women's participation in athletic competitions, and in 1922 co-founded the Women's Amateur Athletic Association in Great Britain. She worked to assure women's participation in the Olympic Games, and succeeded in 1928. In 1929, she was involved in an air crash that left her with severe injuries from which she was not expected to recover; she did, but faced some disabilities from that accident for the rest of her life. She did not solo in an airplane again for almost two years. In the 1930's she was involved in a bitter divorce dispute with Lord Heath, her second husband, which captured media attention with suits and countersuits over unpaid bills. Soon after the divorce, she married George Anthony Smith, whom she taught to pilot. She died of injuries suffered in a fall down the steps of a double-decker bus.

Johnson, Amy (1903-1941): Daughter of a Hull fish merchant, Amy Johnson had to work to support herself and learn to fly in her spare time. She became friends with Pauline Gower and Dorothy Spicer during the late 1920s at Stag Lane Aerodrome where they all worked on flying. In 1929 Johnson became the first woman in the world to earn a ground engineer's license. In 1930, scarcely a year after earning her pilot's license, she embarked on a record solo flight from Croydon, England, to Darwin, Australia, actually setting two records: the speed record for the England to Karachi leg of the flight and the first woman to solo for the entire trip. (Among the benefits of that flight was a free shopping tour at Selfridge's in London, a store owned by Violette de Sibour's father). The women's record she set on that flight stood until 1934 when it was broken by Jean Batten. Johnson's flight earned her the media title of "England's Lindbergh," and along with media popularity, she went on to win a series of records for long distance flights. In 1932 she married Jim Mollison, aviator in his own right, and began a stormy, competitive relationship with him. She was usually considered the better aviator of the two, but their marriage was punctuated by efforts to break each other's records or to set new ones. They divorced in 1938 after having been separated for some time. In 1934, Johnson became the first woman to pilot on a cross-Channel air service, but, she tells us in her autobiography, she was not paid for her efforts and resigned after three weeks. In 1940, she joined the Air Transport Auxiliary under the command of Pauline Gower, where she hoped to help convince the world that women had a real place in aviation. She crashed while flying a mission over the Thames Estuary in 1941, her body never recovered. While rumors of her involvement in espionage for the British government caught public attention, they had little credibility. Although 1941 is listed virtually everywhere as her death date, she was not officially declared dead until 1943.

Lindbergh, Anne Morrow (1906-present): Educated in the classical tradition at Smith College, she married Charles Lindbergh in 1929 and began a career that combined aviation with writing. Always shy by nature, she found the publicity surrounding herself and Charles to be overwhelming at times. Often thought of as a symbol of the new age, the Lindberghs found themselves virtually the prey of the media. The circus-like news coverage that surrounded the kidnapping and murder of their son in 1932 proved to be more than they could endure. By the mid-1930s in an effort to protect their children (two daughters and three sons) from the violence they found in the United States, they had moved to an island off England and toured Europe rather extensively. At the request of the United States government, Charles met with German Nazi officials and toured their airplane factories before he and Anne returned to the United States in 1939. His conclusion that the Germans were far more advanced than the United States in aviation technology led, in part, to his becoming an unpopular activist against the approaching war and to his seemingly racist views. Anne Lindbergh wrote extensively in the early 1950s and as the result of some of her essays about him is largely to be credited with restoring his reputation. After World War II, she began publishing the multiple volumes of her diaries and letters, as well as extended essays such as *Gift from the Sea* (1955; updated 1975) and *Earth Shine* (1966). The grace of her language and her poetic view of the world have made her more renowned today as writer than as aviator.

Murray, Stella Wolfe (?-?): Beyond what she herself told us, not much information exists about her. An invalid from birth, she became the quintessential observer of life. Fascinated by aviation, she turned her views of it into a career of writing.

Spicer, Dorothy (1908-1946): The loss Spicer referred to in the Epilogue of Pauline Gower's book was the death in the winter of 1936-1937 of her father and of Pauline Gower's mother. Although their business relationship ended, the two continued to be active in aviation and related activities. As a member of the Women's Engineering Association, Spicer presented a paper to the group in 1937 on the suitability of steel in the design and manufacture of airplane engines. In 1938, she became a staff member of the Air Registration Board and also obtained her Glider Engineers license. In April of that year she also married Richard Pearse, with Pauline Gower serving as one of her attendants. Her daughter Patricia was born in 1939. Spicer became the London area District Commissioner for the Civil Air Guard and during the war accepted a position from the Ministry of Aircraft Production, where she worked in research and development of more efficient aircraft engines. She became the first woman engineer to be given charge over a fleet of aircraft. She and Richard were killed in an airplane crash over South America, flying as passengers.

Thaden, Louise (1905-1979): Born in Kansas, Louise McPhetridge Thaden began flying at the age of 21. Moving to California, she found herself in the midst of the growing aviation industry. She demonstrated aircraft and set a number of endurance and altitude records. In fact, at one point, she held three international records at the same time for speed, endurance, and altitude. She married aviation engineer Herbert Thaden and became his business partner. In the early 1930s, she co-founded with Amelia Earhart and Ruth Nichols the Ninety-Nines. During the Depression she worked with the U. S. Government to encourage local governments to begin aerial marking systems for the aid of pilots flying over unfamiliar territory. She had two children, one son and one daughter.

Bibliography

Bacon, Gertrude. *All About Flying*. London: Methuen & Co., Ltd., 1915, 1919.

________. *Memories of Land and Sky*. London: Methuen & Co., Ltd., 1928.

Batten, Jean. *My Life*. London: George G. Harrap & Co., Ltd., 1938.

________. *Solo Flight*. Australia: Jackson & O'Sullivan, Ltd., 1934.

Beale, Marie. *Flight Into America's Past*. New York: G. P. Putnam's Sons, 1932.

________. *The Modern Magic Carpet*. Baltimore: J. H. Furst Co., 1930.

Bruce, the Hon. Mrs. Victor. *The Bluebird's Flight*. London: Chapman & Hall, Ltd., 1931.

Camac, Harriet J. M. *From India to England by Air*. New York: Privately Printed, 1929.

De Sibour, Violette. *Flying Gypsies*. New York: G. P. Putnam's Sons, 1930.

Earhart, Amelia. *The Fun of It*. New York: Brewer, Warren & Putnam, 1932.

________. *Last Flight*. New York: Harcourt, Brace & Co., 1937.

________. *20 Hrs. 40 Min*. New York: Grosset & Dunlap, Publishers, 1928, (Published by Arrangement with G. P. Putnam's Sons).

Gower, Pauline. *Women With Wings*. London: John Long, Ltd., 1938.

Heath, Lady and Stella Wolfe Murray. *Woman and Flying*. John Long, Ltd., 1929.

Johnson, Amy. "Myself When Young." in *Myself When Young, by Famous Women of To-day*, ed. Margot Oxford, the Countess of Oxford and Asquith. London: Frederick Muller, Ltd., 1938. 131-156.

________. *Sky Roads of the World*. London and Edinburgh: W. and R. Chambers, Ltd., 1939.

Lindbergh, Anne Morrow. *Bring Me A Unicorn*. New York: Harcourt Brace Jovanovich, Inc., 1938.

________. *Listen! the Wind*. New York: Harcourt, Brace & Co., 1938.

________. *North to the Orient*. New York: Harcourt, Brace & Co., 1935.
________.*War Within and Without*. New York: Harcourt Brace Jovanovich, Inc., 1980.
Thaden, Louise. *High, Wide and Frightened*. New York: Stackpole Sons, 1938.

Index